Introduction to Organic and Medicinal Chemistry

This text is designed to be an undergraduate reference textbook for students of chemistry who aim to select degree modules geared towards medicinal chemistry. The first two chapters offer an overview of basic organic chemistry, followed by organic synthesis in Chapter 3, with an emphasis on the importance of synthetic procedures in relation to the chemistry of drug design. In Chapter 4, the book covers foundational aspects of biochemistry and biomedical science to provide a basis for students understanding *where* and *how* drugs work on selected targets. Later, the book explains how medicines have been created for selected topics in medicinal chemistry – namely against pathogens, cancer and neurological targets.

Features:

- User-friendly text that deals with key methods in organic synthesis and medicinal chemistry
- An overture of drug design against disease targets
- An introduction to the molecular basis of disease and key topics in medicinal chemistry
- Geared towards chemistry undergraduates with an interest in pursuing medicinal chemistry further into their degree
- A complete set of tutorial questions and answers for each chapter

Nathan Keighley studied chemistry at the University of York, with selected option modules in medicinal chemistry and organic synthesis. It was this course content that inspired the writing of this textbook, *Introduction to Organic and Medicinal Chemistry* because he hoped to pass on his interest in the subject to current students of organic and medicinal chemistry. Nathan then focused his studies into organic chemistry with a PhD in Environmental Chemistry, researching how organic molecules used in agriculture biodegrade in the environment to mitigate pollution, and later applied this skillset into a career in industry. Chemistry is a highly relevant subject to modern life and Nathan wishes the students who use this book the best of luck with their exams and hopes that they too are inspired to work in this most important of scientific fields.

Introduction to Organic and Medicinal Chemistry

Nathan Keighley

CRC Press
Taylor & Francis Group
Boca Raton London New York

CRC Press is an imprint of the
Taylor & Francis Group, an **informa** business

Designed cover image: shutterstock

First edition published 2025
by CRC Press
2385 NW Executive Center Drive, Suite 320, Boca Raton FL 33431

and by CRC Press
4 Park Square, Milton Park, Abingdon, Oxon, OX14 4RN

CRC Press is an imprint of Taylor & Francis Group, LLC

© 2025 Nathan Keighley

ISBN: 9781032850429 (hbk)
ISBN: 9781032851563 (pbk)
ISBN: 9781003516835 (ebk)

DOI: 10.1201/9781003516835

Typeset in Times
by codeMantra

Contents

Preface

The subject of chemistry is widely acknowledged as being conceptually challenging and is regarded with a perceived difficulty. Consequently, at school, many individuals are put off by chemistry and tend to avoid it in later education. This is unfortunate because chemistry is a beautiful, interesting and exceedingly important and relevant subject, playing an essential role in society. This textbook aims to enlighten the reader to how important organic chemistry is in terms of medicine. Of course, chemistry's role in society is by no means limited to medicine: materials, food and many technologies rely on chemists and their passion and intrigue for the subject, without which there would be severe limitations to the advancements of these technologies.

At its core, chemistry is a rigorous and systematic subject. This is very much reflected in the nature of chemistry textbooks used in education. In accordance, this textbook has been meticulously well structured to ensure that the content is clearly laid out to facilitate learning. The key elements of each chapter are highlighted in numbered subchapters for easy reference for the student, enabling them to instantly arrive at the required material.

In many contemporary chemistry textbooks, the interesting aspects of the subject matter may be lost. This textbook aims to address this dilemma by producing a text that is a joy to read and makes the subject interesting while still including the foundational principles of chemistry, and not lose the systematic rigour of the subject. This is achieved by having longer written appraisals of selected topics in medicinal chemistry, so that the student can gain a fuller understanding of the application of organic chemistry to medicine and the pharmaceuticals industry.

The core focus of the textbook is nevertheless to be an educational resource to accompany university chemistry degree programmes. The content, as such, has been carefully planned out to achieve this aim. The first chapter is a foundational introduction to organic chemistry to bridge the gap between college and university before moving on to the intricate details of organic synthesis in Chapter 2. This should prepare the student to apply organic reactions to produce synthetic strategies for making target organic compounds. This is in essence the process that medicinal chemists use during the process of drug discovery. The later chapters in this textbook introduce these concepts to students in an applied manner to existing pharmaceuticals.

The systematic nature of chemistry, for me, made it a delight to learn because my studies had structure. In particular, I found organic chemistry to be fascinating, and the main goal for this textbook was to convey my enthusiasm for organic chemistry as well as highlight its fundamental importance in medicine. *Introduction to Organic and Medicinal Chemistry* aims to be an introductory text, so that readers with a basic background education in chemistry can access and enjoy the material and gain an appreciation for the subject. The scope of the textbook is primarily for undergraduate chemistry students who will be learning organic chemistry in their core syllabus and may want to specialise their studies towards medicinal chemistry. This textbook will therefore continue to be a useful resource towards the latter stages of a chemistry degree. Moreover, the accessibility of the textbook hopes to facilitate the studies of non-chemistry students who require a foundation level of knowledge of organic chemistry and medicinal chemistry. This may include students of biochemistry and biomedical science, and perhaps it could serve as a resource for student doctors.

1 Introduction to Organic Chemistry

ABSTRACT

Carbon is in Group 4 of the periodic table, signifying that it has four valence electrons, which it uses to typically form four covalent bonds with other carbon atoms or heteroatoms. This gives rise to the possibility of various organic molecular structures. These may be simple hydrocarbons, such as alkanes and alkenes, formed from chains of carbon atoms, or the formation of carbon ring structures. Where bonding to heteroatoms is involved, characteristic functional groups are formed, which have their own distinctive reactivity. Sometimes, organic molecules may contain the same number of each type of atom, and yet be entirely different molecules. These are called isomers. Analytical techniques exist that can be used to isolate and characterise organic molecules and thus distinguish between the huge variety of organic molecules that might be present in a reaction mixture or a natural product.

Life is based on carbon. Organic chemistry is dedicated to this element, whose properties are defined by the nature of the carbon atom. Carbon has the ability to form four covalent bonds. These are the strong bonds that arise from the sharing of pairs of electrons. In this way, carbon atoms can form the molecular structures that are essential for life on Earth. Indeed, organic chemistry as a subject matter originated as a tentative effort to understand the chemistry of life. From its humble beginnings in the early part of the 19th century the subject has grown extraordinarily, as chemists have studied, researched and understood in greater detail how carbon-based molecular structures react and interact with one another. This has led to the development of a plethora of chemical technologies.

Our modern society is now irrevocably dependent on organic chemicals. Organic chemistry is now fundamental in feeding, clothing and curing the population of illness. The creation of new molecules is invaluable for many, many industries. Materials, such as plastics and other polymer technologies, are so diverse that they can vary in their applications, ranging from clothing, disposable bags, as well as heavy duty structural applications used in aerospace technology. These applications all depend on the precise molecular structure of the materials, which governs their properties. Small organic molecules are used to make dyes for colouring our clothes, to make perfumes, to make food preservatives and endless other applications.

Of particular interest for the scope of this textbook is the use of organic chemistry in medicine. The development of medicine, along with improved sanitary conditions, has been responsible for the dramatic increase in life expectancy observed since the start of the 19th century to the present day. However, natural organic products have been used for curing diseases since the 16th century. One famous example is quinine,

DOI: 10.1201/9781003516835-1

which was extracted from the bark of the South American cinchona tree, and used to treat fevers, in particular malaria. The Jesuits who did this work, of course, did not know the molecular structure of quinine, but merely understood that the substance had medicinally beneficial effects. In many instances, substances have been identified for their medically beneficial effects and isolated as curative agents. This has been the basis for the development of medicine from natural products and continues to this day. However, with advancements in our understanding of the chemical structures of these medicinally useful molecules, medicinal chemists have been able to modify the molecular structures of these compounds to create more effective medicines. The structure of quinine has since inspired the design of modern drugs that are much more effective in treating malaria than quinine itself.

1.1 STRUCTURE AND BONDING IN ORGANIC CHEMISTRY

Carbon is in Group 4 of the periodic table, which identifies that there are four valence electrons. These are the outer electrons involved in bonding, so a further four electrons are required through bonding to other atoms to satisfy the octet rule; hence carbon atoms characteristically form four covalent bonds; each bond comprises shared pairs of electrons with partner atoms. These may be single bonds, with one shared pair of electrons, or double bonds and even triple bonds, with two and three pairs of electrons being shared. It is common for carbon atoms to bond together in chains and rings to produce the carbon skeleton that defines an organic molecule; the remaining valences are often satisfied by hydrogen atoms, which offer their single electron to form a bonding pair. This produces the simplest class of organic compounds: hydrocarbons, typically separated according to molecular mass through fractional distillation of crude oil.

Other elements from the periodic table, generalised as heteroatoms, are also of importance in organic chemistry. Particularly nitrogen, oxygen and the halogens (Group 7) and even organometallic species are all vital components of organic chemistry. To fulfil the octet rule, nitrogen, with five valence electrons, requires three more electrons, so typically forms three bonds; oxygen (Group 6) needs two electrons, so tends to form two bonds; while the halogens require one more electron, so they form single bonds. The valence electrons govern the chemistry of the elements, and so are the most important. However, they do not represent the entirety of the atom's electron configuration; it is worth noting that the total number of electrons, whose charge is fundamentally negative, equals the atomic number, Z, which is the number of positively charged protons within the atomic nucleus to give a neutral atom. The remainder of the atomic mass, which for carbon is 12 g mol^{-1}, is made up of neutrons, i.e. the carbon nucleus is composed of six protons and six neutrons. The number of neutrons may vary to give isotopes, which have a different atomic mass, but the same chemical properties, since it is the valence electrons that govern chemistry. The appropriate number of valence electrons may be added/subtracted to give charged atoms, called ions.

The number of covalent bonds that an atom can form governs molecular geometry. This is explained by valence shell electron pair repulsion (VSEPR) theory, which predicts that pairs of electrons will occupy positions around the atom as far apart from one another as possible. This leads to characteristic molecular shapes. For a carbon atom within a molecule, with four equivalent bonding pairs of electrons, each bonding pair will be positioned equidistantly in three-dimensional space to produce a tetrahedral geometry, with equal 109.5-degree angles between bonds. For a boron atom forming three equivalent bonds, the bonding pairs of electrons will be separated the maximum distance apart to produce a trigonal planar shape, with 120-degree bond angles. Scenarios where multiple bonding is present result in different molecular shapes. If the carbon atom were to have a double bond to one of its neighbours, as in the case of ethene, there will be three points of electron density and consequently the molecule will adopt a trigonal planar shape. However, the electron density is not evenly distributed around the carbon atom; more of it is present in the double bond than in the two single bonds, therefore the shape will be slightly distorted. VSEPR theory predicts that electrons that are not involved in bonding, or 'lone pairs', will repel more and influence the molecular shape. For example, although a nitrogen may have three bonds similar to boron, the presence of a lone pair pushes the bonding electron pairs away to produce a trigonal pyramidal shape. Likewise, water molecules are not linear; the presence of two lone pairs pushes the oxygen–hydrogen bonds down into a V-shape. The shape of molecules, dependent on the nature of the atoms and their bonding, has a dramatic effect on chemistry because molecular geometry influences how reactions proceed.

1.1.1 COVALENT BONDING

Atoms may form **covalent bonds**, where there is a shared pair of electrons in order to attain a full outer shell. This forms a covalent compound comprising molecules. The electrostatic attraction between the shared electrons and nuclei holds these bonds together. In the example given in Figure 1.1, two chlorine atoms each share a pair of electrons to form the diatomic molecule. More than one pair of electrons can be shared during covalent bonding to form double and even triple bonds.

In some cases, an atom with a lone pair of electrons will donate them to an electron-deficient atom to form a **coordinate bond**. This type of bonding is demonstrated by the ammonium ion (Figure 1.2) where the lone pair of electrons on the nitrogen is donated to a proton, giving the molecule a positive charge.

The shared pairs of electrons in a covalent bond are not always evenly distributed. This depends on the electronegativity of the atoms involved.

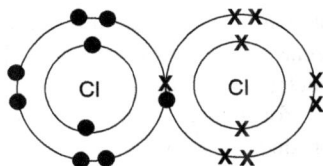

FIGURE 1.1 The covalent bonding displayed between two chlorine atoms, as a simple cartoon representation to illustrate the shared bonding pair of electrons.

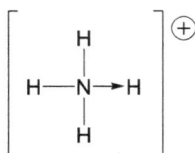

FIGURE 1.2 Example of coordinate bonding in the ammonium ion. Nitrogen forms four covalent bonds to four hydrogen atoms, one of which is via donation of its lone pair of electrons, indicated here with an arrow.

Electronegativity is the power of an atom to attract the electron density in a covalent bond towards itself.

Electronegativity depends on the nuclear charge, atomic radius and the effect of shielding. Across a period, nuclear charge increases, and because there are no additional electron shells to cause shielding, the atoms get smaller and electronegativity increases. Up a group of the periodic table, atoms are smaller, with fewer electron shells, hence the most electronegative atoms are found in the top-right of the periodic table: nitrogen, fluorine and oxygen.

Bond polarity is a property that arises in a covalent bond as a consequence of the atoms within the molecule having very different values of electronegativity. This results in a distortion of the electron density in the bond towards the more electronegative atom.

Although covalent bonds are very strong, the intermolecular forces between **simple molecules** are very weak. This is why simple molecules have low boiling points and many, such as Cl_2, exist as gases. Because these molecules are neutral, they do not conduct electricity. **Macromolecules** consist entirely of covalent bonds, so have very high melting points. Their bonding and structure determine their ability to conduct electricity.

Macromolecular crystals are formed from covalently bonded atoms arranged in a lattice. Because the covalent bonds extend throughout the lattice, macromolecular crystals have high melting points. A lot of energy is needed to break the covalent bonds during sublimation so compounds such as silicon dioxide and diamond, which are macromolecular crystals, have high melting points.

Structure and bonding are fundamental to the properties of the allotropes of carbon. Diamond consists of carbon atoms that form four single covalent bonds with other carbon atoms to produce a tetrahedral shape that is identically repeated throughout the symmetrical structure. This structure enables external forces to be spread throughout the lattice, giving diamond its hardness. Diamond has a high melting point because a lot of energy is needed to break all four covalent bonds. Like most macromolecules, diamond does not conduct electricity because there are no free charged particles.

Carbon atoms in graphite form three single covalent bonds to form a trigonal planar arrangement. This produces a two-dimensional layer of linked hexagons in a honeycomb structure. Each carbon atom has a 'spare' electron in p-orbitals which merge to form a delocalised system above and below the planar layers of the carbon sheets. It is this inimitable feature that enables graphite to conduct electricity along the hexagonal planes. These layers of carbon atoms are held together by weak Van der Waals forces, giving graphite its soft and flaky property as the sheets can slide past one another. Being a macromolecule, graphite has a high melting point, although not as high as diamond.

By contrast, small molecular structures that are based on carbon frameworks have low melting points. Hydrocarbons can form solid waxes, oils and be gaseous at room temperature. Their state depends on the length of the parent carbon chain because this, in turn, governs the number of non-covalent, or intermolecular, forces between molecules. The longer the hydrocarbon chain, the higher the boiling point of the substance. This is because more Van der Waals forces are formed along the longer molecular chains. Therefore, oils tend to be made up of long chain hydrocarbons including decane and higher M_r hydrocarbons, while short chain hydrocarbons, such as methane and ethane, are volatile gases.

1.1.2 Intermolecular Forces

Covalently bonded compounds may exist as simple molecules, where forces between the molecules hold the substance together, as molecules interact with one another via non-covalent interactions. These forces are electrostatic interactions that arise from the uneven distribution of electron density in the molecules. The magnitude of these forces may vary, depending on the properties of the covalent bonds and the electronegativity of the atoms involved in the bonding. However, these interactions are always weaker than covalent bonds, so simple molecules have relatively low melting and boiling points.

Bond polarity arises when atoms in a molecule have different electronegativity. If the molecule is symmetrical in shape, then these effects cancel out, for example, in tetrachloromethane (Figure 1.3). However, if atoms with different electronegativity are not evenly distributed, then this produces a permanent dipole. The electronegative atoms draw electrons towards themselves, creating a partially negative charge to this side of the molecule, while depriving the atoms on the opposite side of the molecules of their electrons, which leads to a partially positive charge. **Dipole–dipole forces** act between molecules with a permanent dipole and the molecules may flip their orientation to attract each other between opposite dipoles. For example, in dichloromethane (also Figure 1.3) the chlorine atoms pull electron density from the carbon–hydrogen bonds, creating a dipole. The opposite poles attract. This means that dichloromethane will have a higher boiling point than tetrachloromethane.

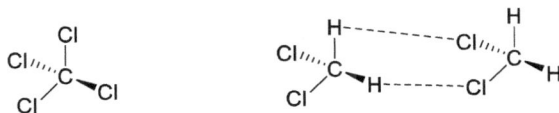

FIGURE 1.3 The displayed structures of tetrachloromethane and dichloromethane, illustrating how the attraction between dipoles can occur. The figure shows non-covalent bonds occurring between chlorine atoms of the hydrogen atoms of neighbouring CCl_2H_2 molecules.

Molecules, such as tetrachloromethane, which do not exhibit dipole–dipole inter-actions, do still experience intermolecular forces. The positive and negative charge that makes up a neutral atom produce very weak electrostatic attractions between molecules as a result of the random instantaneous uneven distribution of electron density around the molecules. These are called **Van der Waals forces**. At any given moment, there could be an uneven distribution of electron density causing a tempo-rary dipole and these instantaneous dipoles induce dipoles in neighbouring atoms. Van der Waals forces work in combination with other intermolecular forces.

The strength of Van der Waals forces increases with the number of electrons pres-ent. This explains why the boiling points of the noble gases increase down the group and the boiling points of hydrocarbons increase with chain length. As the number of electrons present in the molecules increases, there is a greater number of instanta-neous dipoles that can be produced.

Hydrogen bonding is the strongest type of intermolecular force. For hydrogen bonding to occur, a hydrogen atom must be covalently bonded to another atom with a large electronegativity and a lone pair of electrons. Only oxygen, nitrogen and fluo-rine are sufficiently electronegative.

The way in which hydrogen bonding works is as follows: an electronegative atom (N, O or F) withdraws the shared pair of electrons in the covalent bond from the hydrogen and exposes the proton, causing the hydrogen atom to have a partial posi-tive charge. The lone pair of electrons on N, O or F of neighbouring molecules is attracted to the positive charge of the hydrogen. An example of how this bonding occurs in water is shown in Figure 1.4. Hydrogen-bonded molecules tend to have higher boiling points, as more energy is required to break these bonds and separate the molecules, compared with other intermolecular forces.

In its liquid state, water molecules are free to move, and the hydrogen bonds con-tinually break and re-form. However, when frozen, the molecules arrange into a static three-dimensional structure held together by hydrogen bonds. The water molecules must adopt a precise orientation to fit into this lattice, and so are less tightly packed; hence ice is less dense than water.

The forces that exist between molecules have different strengths, depending on the difference in electronegativity of the covalently bonded atoms in the molecules. All molecules exhibit some kind of intermolecular interaction, even if these are just weak Van der Waals forces arising from the instantaneous dipoles that occur across bonds between atoms with similar electronegativity. For covalent bonds between atoms with a large difference in electronegativity, where bond polarity arises, intermolecular

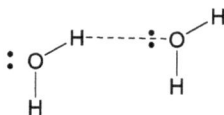

FIGURE 1.4 Displayed structure of water molecules exhibiting hydrogen bonding. The fig-ure shows non-covalent bonds occurring between the lone pair of electrons on oxygen and the hydrogen atoms of neighbouring water molecules.

interactions are stronger because of permanent dipole–dipole forces that exist between molecules. These molecules tend to have a higher boiling point. For the most electronegative atoms, N, O and F, hydrogen bonding can be in place when these atoms are bonded to hydrogen. This is the strongest of the intermolecular forces.

Hydrogen bonds display directionality, hence molecular orientation is important during bond formation, as the N, O or F lone pair of electrons must be aligned with the polarised hydrogen atom in order for bond formation to occur. This applies to intermolecular interactions as well as intramolecular forces between different groups present on a molecule. This directionality of hydrogen bonding is of critical importance to its role in biochemistry and drug binding interactions.

Hydrogen bonding is a crucial feature of supramolecular chemistry and is a prerequisite for the binding of neutral guest compounds to macromolecular structures. The directionality of hydrogen bonding requires binding groups to be preorganised and rigidly held in position within the host binding site. Optimising the number of hydrogen bond interactions is deterministic of the strength of the binding interaction between host and guest molecules, for example, a drug compound and its binding site within the cell. It is also important to consider secondary (diagonal) interactions. Like charges will repel one another; in order to be constructive, hydrogen bond donor and acceptor groups must alternate on opposite sides of the binding site to ensure that secondary diagonal relationships are attractive, therefore strengthening the array. It is therefore important to understand these interactions for designing drugs to fit a receptor, or for applications in nanotechnology. The directional nature of hydrogen bonds makes them well suited for assembly of complex supermolecules, depending on appropriate symmetries and complementarities of hydrogen bond donors and acceptors. The stronger the interactions, the larger the molecular assembly possible. Polymer chemistry utilises these properties of hydrogen bonds. Polymers can be synthesised from monomers containing hydrogen bond donors and acceptors, e.g. nylon and Kevlar are based on polyamides. Hydrogen bonds exist between the polymer chains, making these materials very strong and rigid.

Macromolecular biomolecules, natural polymers, also depend on hydrogen bonding. Hydrogen bonding is of the utmost importance in proteins. Proteins are natural polyamides, comprising amino acid units. The nature of the hydrogen bonding determines the ultimate three-dimensional shape of the proteins, which is essential to the functioning of the protein within its role in the body. For example, a globular protein, such as haemoglobin, has to be perfectly preorganised to contain a haem group for oxygen binding, while a fibrous protein like collagen needs to be strong for structural purposes.

The 3D tertiary structure is determined exactly by the particular hydrogen bonding arrangement in the protein. This is most exquisitely demonstrated by enzymes, where the shape of the active site is critical to its function. Precise complementarity of shape and hydrogen bonding interactions between the substrate and the active site are needed for binding. Furthermore, hydrogen bonding is often an essential feature of the catalytic process by distorting the substrate into a reactive conformation and/or stabilising the transition state, thereby the activation energy of the reaction is lowered. Oftentimes, enzymes are selected as potential drug targets on account of their importance in biochemistry.

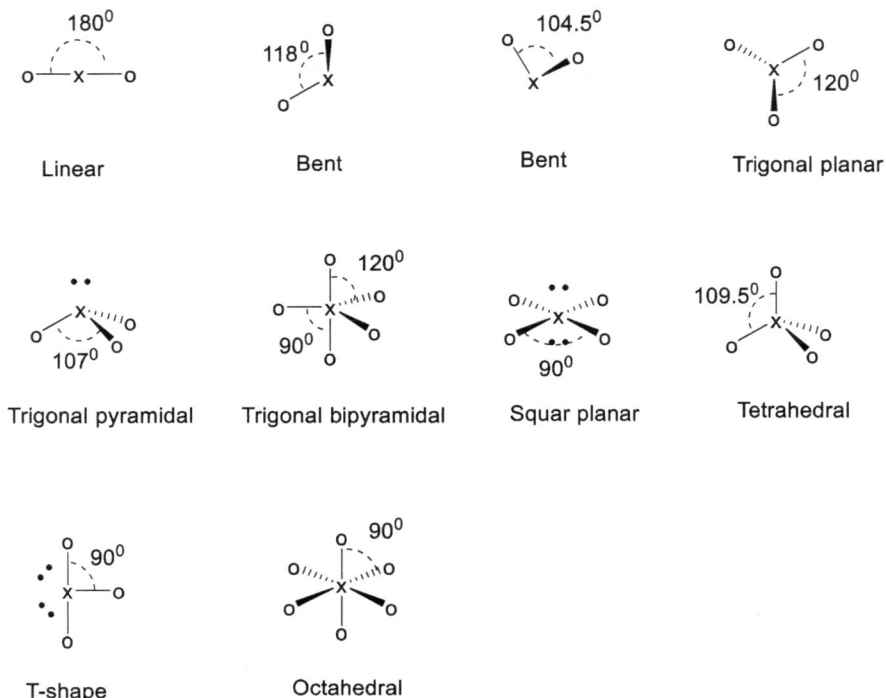

FIGURE 1.5   The geometrical shapes of covalent molecules, according to VSEPR theory, showing bonding angles that result from the repulsion of bonding and non-bonding pairs of electrons.

1.1.3   SHAPES OF MOLECULES

Simple molecules have three-dimensional shapes. **Valence shell electron pair repulsion (VSEPR)** theory suggests that electrons take up positions as far apart as geometrically possible from each other. This leads to predefined molecular shapes. Lone pairs of electrons repel more than bonding pairs, leading to distortions to the normal geometries. These shapes are illustrated in Figure 1.5.

There are some exceptions to the regular bonding patterns. B and Be do not have to have a full outer shell when bonding, P and S can have more than eight electrons in their outer shell because of a vacant 3d-orbital. Atoms above Ca can also have more than eight outer electrons.

It is noteworthy that the atoms in covalent compounds form molecular orbitals formed by the hybridisation of the atomic orbitals. For example, tetrahedral carbon atoms are generated from sp^3 hybridisation of their atomic orbitals. Therefore, none of the bonding orbitals in methane are completely s or completely p. Instead, they are hybrid orbitals that each have partial s character and partial p character. The three 2p-orbitals and the single 2s-orbital hybridise (i.e. mix) to create four hybrid sp^3 orbitals, which are arranged tetrahedrally around the central carbon atom.

1.2 ALKANES

Organic chemistry is dedicated to carbon and the chemistry of this element. Carbon is in Group 4 of the periodic table, so has four valence electrons with which it can form four covalent bonds to other carbon atoms, forming long chains and ring structures, as well as making bonds to other elements, termed heteroatoms, such as oxygen and nitrogen. Often, much of the covalent bonding in organic molecules is between carbon and hydrogen atoms. Carbon atoms can form double and even triple bonds between themselves, producing a plethora of possible configurations.

The structures of organic molecules can be drawn out in **displayed formulas** showing all the atoms and bonding present. However, chemists usually draw **skeletal formulas** to save convolution by representing carbon atoms as the corners of the shapes, omitting the bonds to hydrogen, and only labelling heteroatoms. For notation, the **structural formulas** can be written out. For example, propane would be $CH_3CH_2CH_3$.

Organic compounds can exist in a wide range of structures and carbon chain lengths. Usually, the chemistry of these compounds is governed by the presence of heteroatoms bonded to the carbon chain to form characteristic **functional groups**. These are the reactive sites of the molecules.

*A **homologous series** is a family of organic compounds with the same functional group, but different carbon chain lengths.*

Organic compounds are often derived from crude oil products. Crude oil is the substance created from the breakdown of prehistoric organisms in the Earth's crust over geological time. Crude oil is typically rich in **hydrocarbons**. These are organic molecules that consist of carbon and hydrogen only.

Alkanes are saturated hydrocarbons. They are extracted through **fractional distillation** of crude oil. The crude oil is heated in a furnace and the vaporised mixture is injected into the fractionating tower, which is cooler at the top. The vapours rise up the tower until they reach a fraction that has a lower temperature than their boiling point. The hydrocarbons then condense into a collection tray. The mixture of products in each tray is then piped off. The shorter chain hydrocarbons are nearer the top and the longer chains condense into a residue at the bottom, which is often further processed to yield more products.

Industrial cracking is the process used to split the longer chain hydrocarbons in the residue from fractional distillation into shorter chains, which include more valuable products such as petrol. Additionally, some of the products of industrial cracking include alkenes, which are more reactive than alkanes and are regularly used as chemical feedstocks. There are two main methods of industrial cracking.

1. **Thermal cracking** uses high temperatures (700–1200 K) and high pressures (up to 7000 kPa). The carbon–carbon bonds break to produce two free radicals, which further react to produce an alkene (Figure 1.6).

2. **Catalytic cracking** takes place at lower temperatures (420 K) and lower pressures because a zeolite catalyst is used. These are aluminosilicates that have a honeycomb structure to provide a larger surface area for reactions. They are acidic minerals produced from mining. This method splits branched alkanes to produce motor fuels.

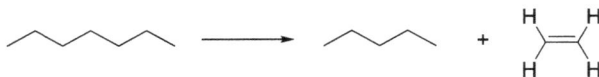

FIGURE 1.6   Reaction scheme for the cracking of the hydrocarbon heptane into an alkane plus an alkene. Here, heptane is cracked into pentane and ethene.

Using alkanes as fuels involves **combustion**. Shorter chain alkanes burn completely producing a lot of heat. Longer chain alkanes usually do not have an adequate oxygen supply to burn completely, and produce other combustion products, such as carbon monoxide and particulates. The combustion products of hydrocarbons used in industry have contributed to **atmospheric pollution**. Such pollutants include SO_2, which reacts in the atmosphere to produce acid rain; CO_2, which is partly responsible for the greenhouse effect and global warming; CO, which is poisonous; NO and H_2O and particulates, which contribute to smog, which can irritate the lungs. Chemical methods to minimise this harmful effect include, for example, flue gas desulphurisation, where calcium oxide, or limestone, is used to absorb the sulphur dioxide in smoke. Another chemical method to reduce pollution is the use of **catalytic converters** in car exhausts. A honeycomb structure is coated in rhodium or platinum which facilitates the catalytic reaction of exhaust gases to produce less harmful products. See the reaction schemes for the combustion of (a) short chain and (b) long chain hydrocarbons. Also, the catalytic reaction schemes (c) and (d) for pollutants are converted into less harmful products in a catalytic converter.

a. $C_2H_{6\,(g)} + 3\frac{1}{2}O2_{(g)} \rightarrow 2CO_{2(g)} + 3H_2O_{(l)}$
b. $C_5H_{12(l)} + 8O_{2(g)} \rightarrow 5CO_{2(g)} + 6H_2O_{(l)}$
c. $2CO_{(g)} + 2NO_{(g)} \rightarrow N_{2(g)} + 2CO_{2(g)}$
d. $C_8H_{18(l)} + 25NO_{(g)} \rightarrow 12\frac{1}{2}N2_{(g)} + 8CO_{2(g)} + 9H_2O$

Carbon chains can branch into different shapes. The hydrocarbons separated from crude oil are not merely composed of linear chains of carbon atoms, but in fact comprise a plethora of branched chain structures as well as cyclic ring structures. This means that hydrocarbons with a generic molecular formula, such as C_5H_{12}, can represent a number of different structures, each with their own chemical characteristics. These are called isomers.

Isomers are molecules that have the same molecular formula, but a different arrangement of atoms.

This kind of structural isomerism involves three categories of isomers: (i) **Positional isomers**, where the functional group is at a different position on the carbon chain; (ii) **Chain isomers**, where the hydrocarbon chain is arranged differently; (iii) **Functional group isomers**, where there are different functional groups possible for the same molecular formula. These are illustrated with examples in Figure 1.7.

In Figure 1.7, the isomers are illustrated as skeletal formulas for clarity. Each corner represents a carbon atom; heteroatoms are clearly shown, and hydrogen atoms are omitted for clarity. This is the most common way that chemical structures are drawn

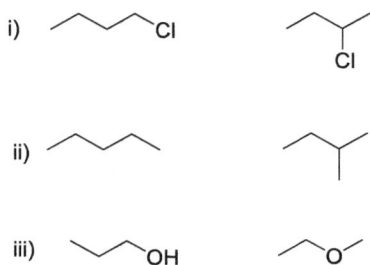

FIGURE 1.7 Examples of pairs of unique molecules that have the same molecular formula, demonstrating (i) positional isomers, (ii) chain isomers and (iii) functional group isomers.

in chemistry. However, for the purpose of learning the basics, some figures later in the text show full bonding. These displayed formulas might also be a requirement in some exam questions.

Hydrocarbon chain isomers are unique molecular structures with their own physicochemical properties. For example, the boiling point of alkanes increases as linear chains become longer, as there are greater numbers of Van der Waals forces between molecules. However, for branched chain isomers with the same molecular formula, the boiling point will be lower because their electron density is more contained within the molecule, so Van der Waals forces are comparatively weaker. Chain isomerism will also have an impact on the reactivity of functional groups attached to the hydrocarbon skeleton. This is because branched chains obstruct the approach of reactant molecules towards the functional group.

Chain isomerism has a simple system for nomenclature. The longest parent chain is acknowledged as the main part of the structure, then the smaller side chains are attributed as prefixes, with a numbering system to account for their position on the parent chain. For the example of chain isomerism in Figure 1.16, the linear chain is pentane, and its isomer has a single carbon atom branching from a four-carbon chain on the second carbon atom of that chain. This isomer can therefore be called 2-methylbutane. However, because there is no other position on the four-carbon chain available for branching, the '2' can be dropped, so that the name becomes methylbutane. There is also a third isomer possible for this molecular formula. A three-carbon chain may have two branches from the second carbon and would be called 2,2-dimethylpropane, or dimethylpropane.

This section has introduced some of the basic concepts of organic chemistry. Carbon atoms can bond to one another to produce a variety of different carbon chain or ring structures. These are the scaffolds of organic molecules. Often heteroatoms will bond to these scaffolds to create the characteristic functional groups in organic chemistry, which display predictable and synthetically useful reactions. These organic structures are represented by a standardised notation system and can be named in accordance with a well-defined system of nomenclature, IUPAC. This brings order to the variety of organic structures that are possible from the nature of isomerism.

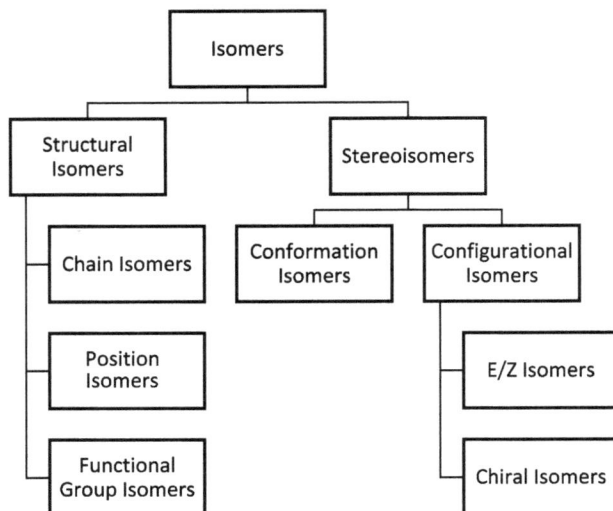

FIGURE 1.8 Flow diagram showing the relationship between different groups of isomers.

1.3 ISOMERISM

Isomers are molecules that have the same molecular formula but have different arrangements of atoms. These may include structural isomerism and stereoisomerism. There are different types of isomerism within each of these groups, as shown in the flow chart, Figure 1.8. Both types of isomerism have important implications for the synthesis of organic molecules and design of novel drug molecules.

1.3.1 STRUCTURAL ISOMERISM

Two organic molecules can be composed of the exact same proportions of atoms, i.e. have the same molecular formula, but the way in which the atoms are arranged may be completely different to give two unique molecules. These are termed **isomers**. There are three different types of structural isomerism. The functional group may be placed at different positions on the carbon chain in positional isomers, or the atoms may be arranged in such a way to give a different functional group, known as functional group isomerism. Equally, it can be the hydrocarbon chain itself that is arranged differently to give different chain isomers.

Figure 1.9 demonstrates how atoms can be arranged to give different molecules from the same molecular formula. Note that the way in which these molecules are drawn is the skeletal formula. Each corner represents a carbon atom; only important heteroatoms are labelled, and the remaining valences of the carbon atoms are bonds to hydrogen atoms (not drawn to save convolution). It can be clearly seen that three of the molecules are alcohols (OH) and molecule (c) is a different functional group. The position of the OH is different between molecules (a) and (b), while in molecule (c), the carbon chain has branched. Each of these molecules will have different properties and reactivity.

(a) [structure] OH (b) [structure with OH] (c) [structure with O] (d) [structure with OH]

FIGURE 1.9 Structural isomers possible for the molecular formula $C_4H_{10}O$.

1.3.2 STEREOISOMERISM

The way in which these molecules are drawn does not portray their three-dimensional structure, which is important because molecules with the same structural formula can be arranged differently in space to produce **stereoisomerism**. Molecules with the same structure can exhibit different configurations, where they may exist as nonsuperimposable mirror images, or groups of atoms may be held in different spatial arrangements either side of a rigid carbon–carbon double bond. This concept may be difficult to envisage, but is crucially important in drug design, as will be seen later in the text. For example, the unfortunate consequences of thalidomide, used for morning sickness, was due to the drug been administered as a 50:50 mixture of the mirror images, where one of the isomers caused harm.

Conformation and configuration are important concepts in organic chemistry. Hence, it is necessary to define what is meant by these terms and the nomenclature used to describe each of them. It is important to understand the connection between conformation and configuration during reaction processes, for example, during addition to alkenes, elimination reaction, ring closing reactions and pericyclic processes, which are all important processes used commonly in chemical synthesis and may, by extension, have implications for the efficacy of potential drug compounds.

Remember that stereoisomers have the same structural formula, but the atoms are arranged differently in space. **Configurational isomers** arise because there is no rotation around a double bond. The nomenclature works by assigning relative priorities to the substituents around the double bond. Highest priority is given to the highest atomic number atom attached to one of the double bond carbon atoms. When the substituents have the same atomic number, which is often the case with alkenes that have long carbon chains, use the first point of difference rule: work back from the double bond until a difference in atomic number is found. Treat multiple bonds as though they were single bonds attached to a ghost atom (e.g. HC=O is carbon bonded to O, O, H). If the two highest priority substituents are on the same side, the alkene is 'Z'; if they are on opposite sides of the C=C double bond it is 'E'. Configurational isomers have different physical and chemical properties, where the E-isomer tends to be more stable due to less steric strain over the double bond. Because the larger substituents are on opposite sides of the double bond, their bulk causes less of a steric hindrance over the confined space of the double bond.

Worked example: Assign the full IUPAC name and corresponding stereochemistry to each of the following two chemical structures in Figure 1.10.

Answer: (Z)-3-chloro-2-methyl-3-phenyl propanoic acid and (E)-1-bromo-2-methoxy-3,4,4-trimethylpent-2-ene

Isomers with chiral centres are asymmetrical and nonidentical with their mirror images, which are therefore nonsuperimposable. These are called **enantiomers**. The sp^3 hybridised tetrahedral carbon atoms that make up organic compounds may

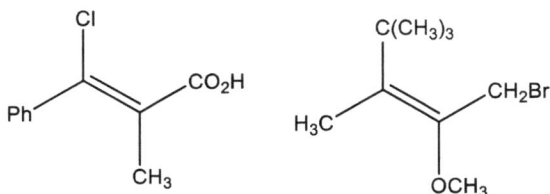

Answer: (Z)-3-chloro-2-methyl-3-phenyl propanoic acid and (E)-1-bromo-2-methoxy-3,4,4-trimethylpent-2-ene

FIGURE 1.10 An example of E/Z isomerism for the nomenclature of two organic compounds.

contain four different groups attached to the 'chiral centre'. The two enantiomers are said to be optically active, as they can rotate plane polarised light either clockwise or anticlockwise, when measured with a polarimeter. This is one method to distinguish enantiomers. In a 50:50 mixture of enantiomers, called a racemate, the effect is cancelled out. Molecules demonstrating chirality exhibit 'handedness' as each mirror image is nonsuperimposable analogous to your right and left hand. The shapes of the mirror images are the same, but your right hand will not fit in your left glove and by analogy, chiral drug molecules will not fit into the same receptor sites, hence stereochemistry is of paramount importance for considering the synthetic approach taken towards designing novel drug compounds.

The R/S nomenclature used to describe enantiomers is determined by using the Cahn–Ingold–Prelog sequencing rules: highest priority is given to the substituent with the highest atomic mass attached to the central chiral carbon or working away from the chiral centre to the first point of difference. Multiple bonds are treated as single bonds attached to 'n' singly bonded atoms. Then, orientate the structure so that the lowest priority group is at the back, then follow the order of priority of the remaining groups with an arrow: clockwise = 'R' and anticlockwise = 'S'. Many organic molecules have more than one chiral centre and each of these needs to be considered during the development of medicinally useful chemicals. For a molecule with n chiral centres, 2^n stereoisomers are possible. Not all of these will be mirror images (enantiomers), these are termed diastereomers. These isomers have different physical and chemical properties, so can be separated. The structures given in Figure 1.11 demonstrate how the Cahn–Ingold–Prelog sequencing rules can be applied.

Conformational isomers result from rotation about a single bond to produce different spatial arrangements of the same compound, which rapidly interconvert. Conformers can be described as staggered or eclipsed to describe the proximity of groups around a single bond. These conformations are each associated with a difference in energy level, which arises from the torsional strain caused by steric effects. When large substituents are in a conformation that brings them close together, the resultant steric hindrance produces torsional strain, which causes the conformation to be a higher energy structure. These conformations are: antiperiplanar, anticlinal, Gauche, and synperiplanar. They describe the relative positions of the substituents around the single bond. Single bonds in rings also show conformational isomerism, as the rings pucker to adopt a lower energy state that favours tetrahedral geometry.

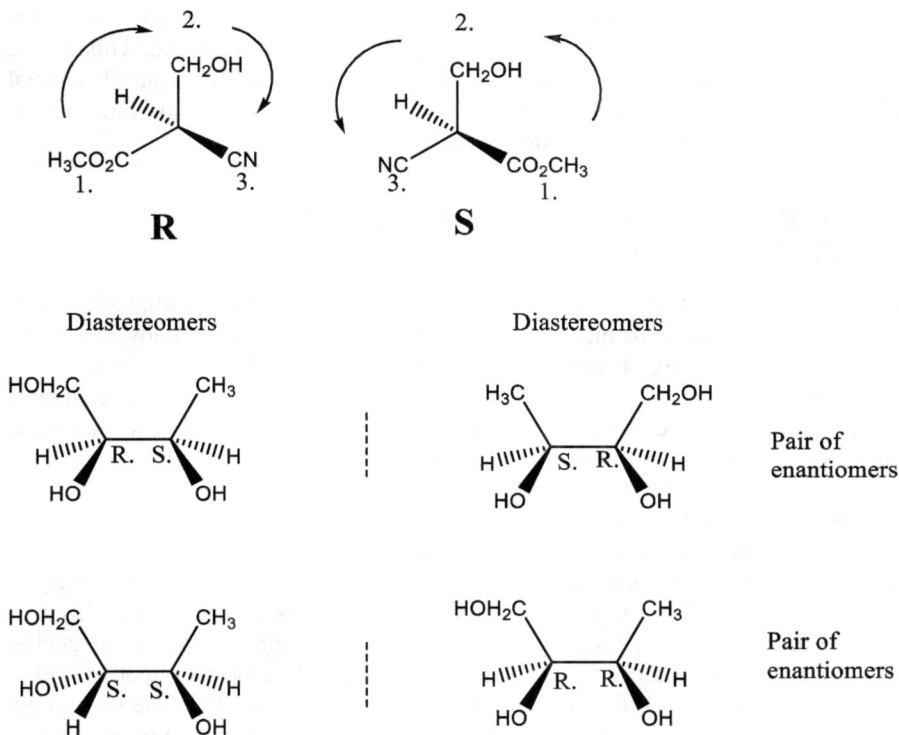

FIGURE 1.11 Diagram of chemical structures to illustrate how the Cahn–Ingold–Prelog sequencing rules of nomenclature can be applied.

Different reaction processes will result in different stereoisomers being produced. For example, Br_2 addition to an alkene leads to an anti-configuration addition product because S_N2 reaction of the Br⁻ ion with the bromonium ion intermediate leads to an inversion of configuration. Elimination reactions of haloalkanes will produce an E-alkene, being the most stable structure due to steric hindrance. The geometry of the alkene starting material in pericyclic reactions influences the stereochemistry of the products.

Organic reactions are used to synthesise drugs. Considerations of the stereochemistry are obviously vital in the design of new medicines. In cases where one of the two stereoisomers is the active drug, an asymmetric synthesis is required where special measures are taken to ensure stereospecificity. With knowledge of the characteristic reactions that different functional groups display, organic chemists can synthesise a target drug molecule from the relevant readily available starting materials. To build a specific target molecule, making carbon–carbon bonds is essential. Functional groups that will undergo addition reactions are useful for this purpose. To ensure strong interactions with the drugs' biological target, a particular functional group may be needed in the molecule. Here a substitution reaction may be relevant. Ultimately, a drug molecule is made with the correct size, shape, correctly positioned functional groups and chemical properties that will interact with the biological system to produce a biological response. Whether the compound acts as a medicine or a poison depends on the dose level of the compound. This can be described by the drug's therapeutic index, which is a measure of a drug's beneficial effect at low dose versus its harmful effects at high dose. No drug is absolutely harmless, and drugs may vary in the side effects they have.

1.4 NOMENCLATURE OF MULTIPLE FUNCTIONAL GROUP COMPOUNDS

Hydrocarbons consist of hydrogen and carbon only. They may be aliphatic, forming chains, or alicyclic, forming an enclosed ring. Saturated compounds, consisting entirely of single bonds, belong to the homologous series of alkanes. Unsaturated compounds, containing multiple carbon–carbon bonds, include alkenes (double bonds with 120-degree bond angles) which can form aromatic rings (benzene) and alkynes (triple bonds; 180-degree bond angles).

Functional groups are the part of a compound that gives it characteristic properties and molecules may be classified on their functional group level, which is the number of bonds to heteroatoms from a given carbon atom. Functional group level can also be referred to as oxidation level and is a key indicator of the reactivity of organic molecules. The highest functional group numbers are the most highly oxidised species; the lowest functional group numbers are the most reduced species. Hence a redox reaction occurs to change one functional group into another with a different oxidation level, for example, the oxidation of an alcohol. The OH group, where carbon has one bond to oxygen, is converted into an aldehyde. Now the carbon atom had two bonds to oxygen and therefore has a higher functional group level, being oxidised. Note that the aldehyde could be further oxidised to a carboxylic acid, where the carbonyl carbon now had three bonds to oxygen atoms.

Organic reactions often involve converting one functional group into another. Therefore, it is possible to have a variety of molecular structures containing different functional groups, which have diverse chemical properties. To add a level of organisation to organic chemistry, it is necessary to catalogue organic compounds using a systematic protocol for their nomenclature (International Union of Pure and Applied Chemistry, IUPAC). This is summarised in Tables 1.1 and 1.2 and the displayed structures of these selected functional group are illustrated in Figure 1.12. Organic compounds with multiple functional groups are named using the following protocol:

- Find the longest alkane chain and name this as the parent alkane.
- Identify the major functional group and replace the -ane suffix.
- Number the chain starting from the functional group.
- Identify any substituents, including any minor functional groups; their names and position number are given in the prefix in alphabetical order.

To name substituted benzenes, the prefixes for the functional groups are listed in alphabetical order, followed by the word 'benzene'; the substituent of lowest alphabetical order is assigned the number 1 position on the ring. Where common names are used, such as benzoic acid, the functional group that it relates to (COOH) is given the number 1 position.

TABLE 1.1

Abbreviations of Alkyl Groups Are Often Used When Writing Formulae

R	Any alkyl group
Me	Methyl -CH_3
Et	Ethyl -CH_2CH_3
P_r	Propyl -$CH_2CH_2CH_3$
^iP_r	Isopropyl -CH_3CHCH_3
^sBu	Sec-butyl $CH_3CHCH_2CH_3$
^iBu	Isobutyl $CH_3CH(CH_3)CH_2$
^tBu	Tert-butyl $CH_3(CH_3)2C-$
Ph	Phenyl C_6H_6
Ar	Any aromatic ring

TABLE 1.2

The Order of Precedence of Different Functional Groups for the Nomenclature of Organic Compounds under the IUPAC System

Highest Priority	Functional Group	Suffix	Prefix
COOH	Carboxylic acid	-oic acid	
COOC	Ester	-oate	Alkoxycarbonyl-
CNHO	Amide	-amide	Amido-
CN	Nitrile	-nitrile	Cyano-
CHO	Aldehyde	-al	Oxo-
COC	Ketone	-one	Oxo-
COH	Alcohol	-ol	Hydroxy-
CNH_2	Amine	-amine	Amino-
CCH_2	Alkene	-ene	Alkenyl-
CCH	Alkyne	-yne	Alkenyl-
CCH_2	Alkane	-ane	Alkyl-
CH_2X	Halogenoalkane		Halo-
Lowest priority			

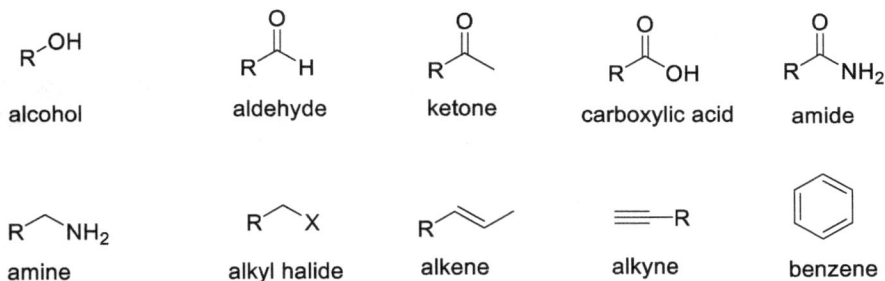

FIGURE 1.12 Displayed structures of selected functional groups where R = the rest of a molecule and X = halogen atom, i.e. F, Cl, Br, I.

1.5 ANALYTICAL TECHNIQUES IN ORGANIC CHEMISTRY

This section is an introduction to analytical techniques used for structure determination. There are a number of instruments widely available in chemistry laboratories that can be used to provide information to assist in the characterisation of chemicals. Analytical chemistry is one of the largest sectors in the chemical industry.

1.5.1 MASS SPECTROMETRY

The molecular weight of organic molecules can be calculated using mass spectrometry. This involves injection of a vaporised sample, which is ionised in the mass spectrometer, and accelerated towards a detector, which registers the ion's mass-to-charge ratio. Sample molecules are bombarded with electrons from an electron gun in the mass spectrometer to cause the removal of one electron to form the **molecular ion**. This is shown as the peak furthest to the right on a mass spectrum. Refer to the representation of a mass spectrum of chloroethane. The molecular ion is the one with the M/Z ratio of 64, corresponding to the molecular weight of chloroethane.

Note that the molecular peak is not the tallest peak in the spectrum. This is known as the **base peak** and corresponds to the most abundant fragment ion. When the sample molecule is bombarded by electrons, the energies involved can lead to **fragmentation**, where the molecular ion is broken up into one fragment with a positive charge and one fragment with an unpaired electron (radical) which is not detected. This is illustrated in the scheme in Figure 1.13.

Sometimes there can be a small peak in the mass spectrum to the right of the molecular ion peak. This is due to the presence of isotopes. In the example given in Figure 1.14 for chloroethene, there is a small peak present to the right of the molecular ion. This is due to the presence of the ^{37}Cl isotope, which only has one quarter of the abundance of the ^{35}Cl isotope.

Fragmentation can also be used to identify the presence of isomers in the sample. Although each molecule has the same molecular weight, each will fragment differently to give different mass spectra. Mass spectrometry can also be used to determine some detail of the nature of the functional groups that are present in the isomers. This is because different functional groups will fragment in predictable ways. For example, carbonyl compounds will typically fracture at the carbonyl group to form an alkyl radical and an acylium ion.

1.5.2 INFRA-RED SPECTROSCOPY

When a sample is placed on an infra-red spectrometer, IR radiation is projected onto the sample of organic molecules. The infra-red radiation consists of a range of frequencies. Frequencies that correspond to the natural vibrational frequencies of the bonds in the organic molecules are absorbed, causing the bond to resonate.

$$\overset{+\bullet}{M} \longrightarrow X^+ + Y^\bullet$$

FIGURE 1.13 Reaction scheme showing a schematic representation of a molecular ion undergoing fragmentation to produce an ion and a radical.

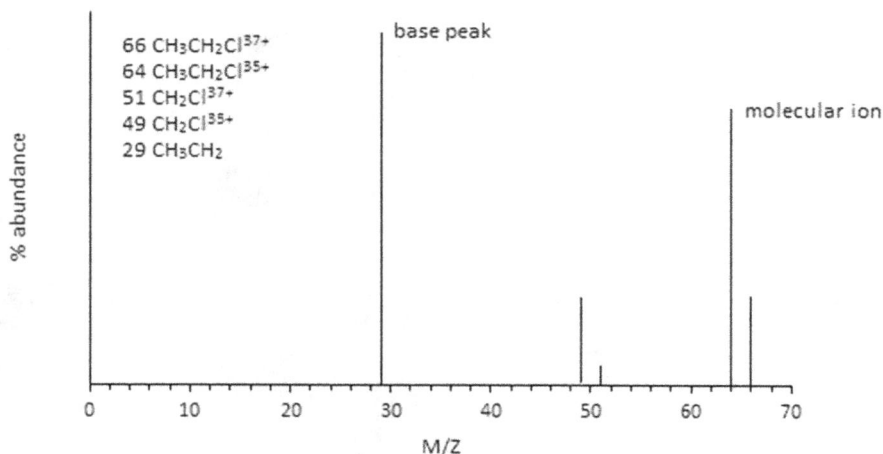

% abundance

66 $CH_3CH_2Cl^{37+}$
64 $CH_3CH_2Cl^{35+}$
51 CH_2Cl^{37+}
49 CH_2Cl^{35+}
29 CH_3CH_2

base peak

molecular ion

0   10   20   30   40   50   60   70

M/Z

FIGURE 1.14 Representation of the mass spectrum of chloroethane, with the base peak and molecular ion peaks labelled.

TABLE 1.3
Sample of Infra-red Absorption Data for Selected Bonds

Bond	Wavenumber (cm⁻¹)
C–H	2850–3300
C–C	750–1100
C=C	1620–1680
C=O	1680–1750
C–O	1000–1300
O–H (alcohols)	3230–3550
O–H (acids)	2500–3000

These frequencies will be missing from the spectrum. Different types of bonds have different resonant frequencies; therefore, these will produce characteristic absorption bands in the IR spectrum. Some examples of these characteristic absorption bands are given in Table 1.3. Functional group signals occur mostly at a frequency above 1500 cm⁻¹ and are usually easy to identify.

The area of an IR spectrum below 1500 cm⁻¹ is known as the **fingerprint region** because this is unique to any particular substance. These signals are produced by all the collective bond vibrations of the various bonds in the molecules that absorb IR radiation. Consequently, this can be used to characterise an unknown substance by comparing it to the spectra of known pure compounds. It is also a means to test for impurities.

In the IR spectrum given in Figure 1.15, the carbonyl peak can be identified at 1649 wavenumbers, which is slightly shifted out of its normal range, being part of an aromatic system. Below 1500 wavenumbers, the spectrum becomes more complicated. This is the fingerprint region, and it will be unique to this compound.

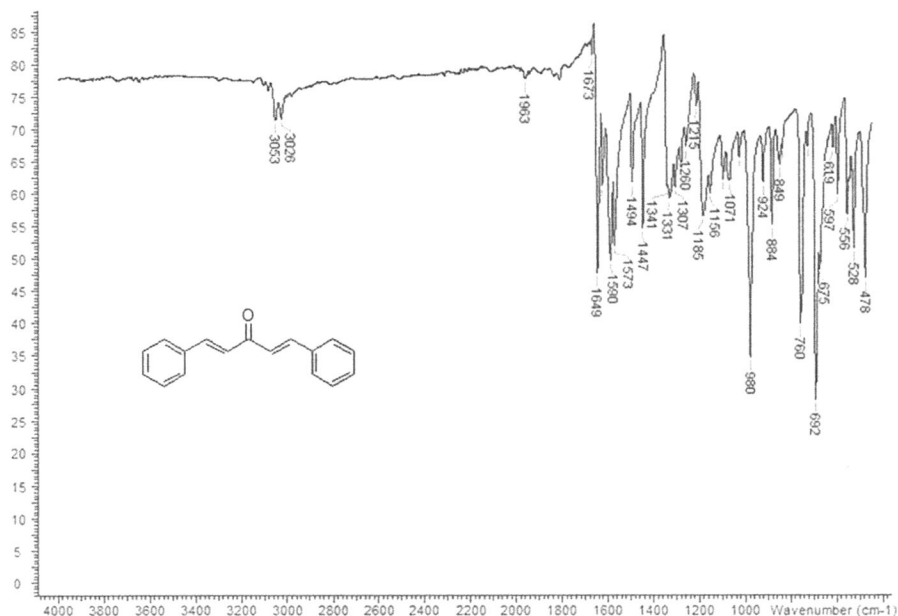

FIGURE 1.15 IR spectrum of dibenzylideneacetone illustrating the carbonyl peak.

IR spectroscopy is a useful analytical technique. The instruments are usually not as expensive as MS or NMR spectrometers and provide a fast and effective analysis of what functional groups might be present in the compound, and the fingerprint region can be used to characterise chemicals against a database of known compounds. However, to achieve a full structural characterisation, NMR spectroscopy is required.

1.5.3 NUCLEAR MAGNETIC RESONANCE SPECTROSCOPY

The basic principles of how nuclear magnetic resonance works is that the instrument uses radio waves to make nuclei in the sample resonate. A magnetic field is applied to the sample to 'freeze' the nuclei in a fixed orientation (based on their quantum spin). This magnetic field is then reversed, and the spins flip, which is associated with a change in energy. This energy change is what is recorded in the NMR spectrum. NMR spectroscopy is most commonly done using ^{13}C and ^{1}H nuclei.

The orbiting electrons of neighbouring nuclei shield the nuclei from the external magnetic field. Because different neighbouring groups are present in a molecule, each group of nuclei will be shielded by different extents and are said to be in different **environments**. Different frequencies of radio waves are therefore required to make the nuclei resonate and therefore the **chemical shift** will be different. Chemical shift is measured relative to a standard, tetramethylsilane (TMS), $Si(CH_3)_4$.

The greater the electron density around a nucleus, the smaller the magnetic field experienced by the nucleus, and the lower the frequency at which it resonates; it is shifted up-field in the NMR spectrum.

The reason that TMS is used to calibrate the NMR spectrum is that it produces only a single peak in its NMR spectrum, as all the ^1H hydrogen atoms and the ^{13}C carbon atoms of the methyl groups are in the same environment because the molecule is symmetrical. The substance is also inert and non-toxic. TMS has a low boiling point, so can easily be removed from the sample.

The number of peaks in a ^1H NMR spectrum is equal to the number of environments in which the nuclei are present. The size of the peaks is proportional to the number of nuclei in that particular environment and is shown by the **integration trace**. These peaks, in fact, split up into smaller peaks due to a process called spin–spin coupling. This occurs as a result of the applied magnetic field on a hydrogen atom being distorted by the magnetic field produced by the nuclei of hydrogen atoms on adjacent carbon atoms. The splitting pattern follows the *N + 1* **rule**.

N hydrogen atoms on an adjacent carbon atom will split the peak into N + 1 smaller peaks.

Some examples of ^1H NMR spectra are given in Figures 1.16 and 1.17. In the ^1H NMR spectrum of iodoethane (Figure 1.16) there are two signals, corresponding to the two environments highlighted on the molecular structure. The signals are split up into a quartet, suggesting that there are three adjacent hydrogen atoms, and a triplet, suggesting two neighbouring hydrogen atoms. In the ^1H NMR spectrum of 1-hydroxypropan-2-one (Figure 1.17) there are three peaks, each a singlet, as there are no adjacent hydrogen atom environments.

It is important to pay attention to the splitting patterns because the peaks can sometimes be misleading. A peak split by one hydrogen atom on adjacent carbon atoms will be a doublet with a 1:1 height ratio. A peak split by two hydrogen atoms on adjacent carbon atoms will produce a triplet, with a height ratio of 1:2:1. A peak split by three hydrogen atoms on adjacent carbon atoms will produce a quartet with a height ratio of 1:3:3:1.

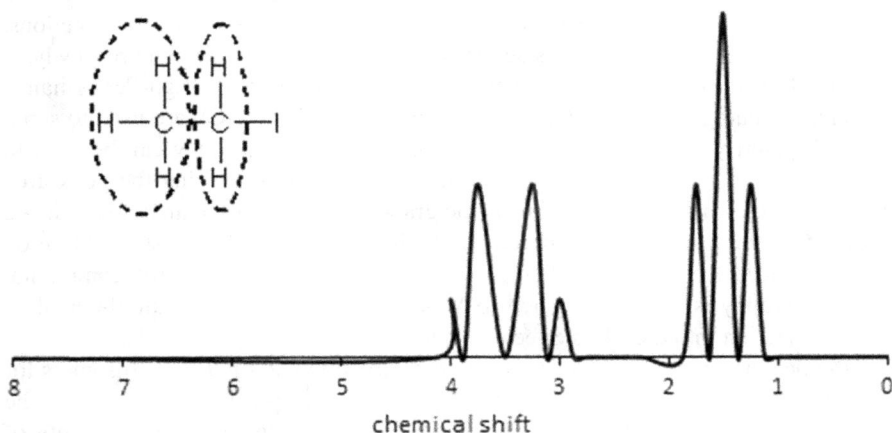

FIGURE 1.16 Representation of the NMR spectrum for iodoethane. The spectrum shows two peaks due to the two environments in which the hydrogen nuclei are present. The peak with the lower chemical shift is split into a triplet and the peak with the higher chemical shift is split into a quadruplet.

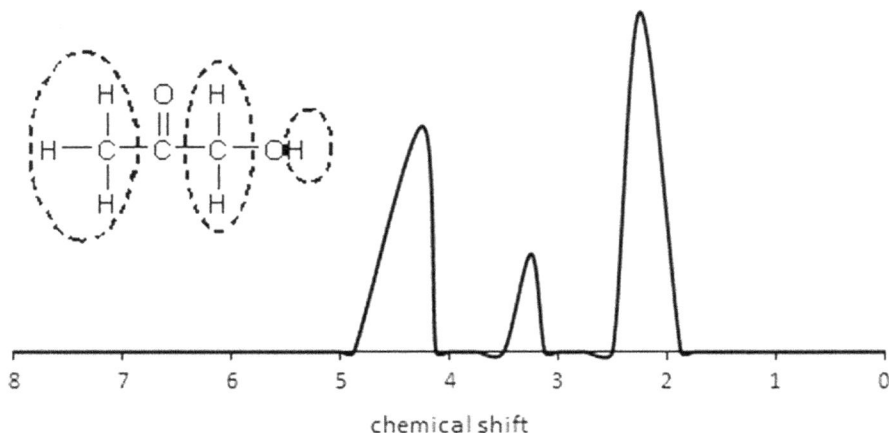

chemical shift

FIGURE 1.17   Representation of the NMR spectrum for hydroxypropanone. The spectrum contains three singlet peaks.

1.5.4 CHROMATOGRAPHY AND SEPARATION TECHNIQUES

Chromatography encompasses a variety of techniques where a mixture, dissolved in a solvent (the **mobile phase**), is passed through a solid **stationary phase** to separate the constituents. The mobile phase carries the soluble components of the mixture with it, as it moves down a column that is packed with the stationary phase. The more soluble these analytes are, the faster they move through the column. The stationary phase interrupts the passage of analytes that have an affinity for it. Therefore, the stronger the interactions between the analyte and the stationary phase, the slower it passes through the column. The column tends to be packed with beads (0.2 µm) to increase the surface area of the stationary phase to maximise binding interactions. Better separation of the mixture is achieved when the analytes travel more slowly.

In analytical chemistry labs, sensitive instruments, such as High-Performance Liquid Chromatography (HPLC) and Gas Chromatography (GC), are used to separate and quantify chemicals with high resolution. Chromatography can be used to measure the concentration of an unknown compound by measuring the peak area produced in the chromatogram. This is the graphical output of the analyte produced by the computer attached to the chromatograph. It is measured against a calibration curve produced from a series of lineated known concentrations of the compound. Chromatography is often conjugated to Mass Spectrometry to separate the analyte prior to injection into the Mass Spectrometer.

Organic compounds are required in high purity, often as single stereoisomers for applications in the pharmaceutical industry. A range of practical methods can be used to purify compounds and the purity can then be assessed using a variety of analytical techniques. Several chromatography techniques are available to purify organic compounds, including HPLC, GC and column chromatography. The column is packed with silica and a solvent is chosen as the mobile phase which dissolves the organic product. The desired product and by-products/unreacted materials will have a different affinity for the silica stationary phase and therefore will have different retention times in the column, so the desired product can be isolated. The degree of separation may not be enough to separate a stereoisomer. HPLC or GC (for volatile compounds) may be needed. In these techniques, the length of the stationary phase is much greater, therefore better separation can be achieved, even isolating one stereoisomer is possible.

When presented with a sample solution containing two (or more) components, it is necessary to select the right solvent system for separation by chromatography. A mixture of two solvents is typically used: one that is polar and the other non-polar, in this case, 85:15 mixture of petroleum:ethyl acetate. The polar solvent, the ethyl acetate, will dissolve the polar component of the crude reaction mixture. This produces a mobile phase that will pass through the silica stationary phase and elute, separating each component of the crude reaction mixture in accordance with its affinity to the stationary phase.

If the analyte has polar groups, such as an alcohol group, with the propensity to form strong hydrogen bonds, it will have affinity for the silica stationary phase, with the tendency to adsorb onto it more strongly than less polar compounds. As a result, the non-polar compounds move with the solvent and pass through the column (or up a TLC plate) with a greater velocity and elute first (smallest retention time). As well as its function in the column, it is also necessary to consider the price of the solvent, as it can be expensive, due to large volumes of waste at the end. Also, it is worth considering the boiling point of the solvent for extraction of the product by rotary evaporation.

The magnitude of the retention times and resolution of the column depend on the dimensions of the column. Ultimately, the size of the column depends on the amount of silica required: a weight ratio of 50 g of silica will suffice for separating 1 g of dissolved product. Making the column longer and thinner would increase retention times and decrease resolution, caused by spreading of the bands over more fractions. Time is also an important consideration when running the column within the constraints of a working day/week: the components of the reaction mixture ideally should be separated within 20 fractions.

Chromatography experiments can be coupled with MS, which enables the molecular mass of the test compound to be accurately determined and therefore allows for the identification of target molecules. However, this may not give information about isomers. IR spectroscopy is a useful and quick technique to identify functional groups present in an analyte, but detailed structural information is often needed. This is where NMR spectroscopy comes in. This enables accurate structural characterisation and information from coupling constants and integration curves can help determine the properties of each stereoisomer.

However, before any of these separations and analytical techniques are pursued, there are a number of practical methods during the synthesis that can be employed to optimise yield and improve the purity of the crude product. For example, it may be necessary to have one reagent in excess, e.g. during the aldol condensation there might be 5 equivalents of aldehyde used to prevent the self-condensation of the ketone. The use of a catalyst may be employed to speed the rate of one reaction over a competing reaction. Equally, the solvent system may impact on this. For an asymmetric reaction, to avoid creating the wrong stereoisomer, a chiral auxiliary can be used, such as oxazolidinones, or via consideration for the choice of nucleophiles used during Cram chelation control.

Once the synthesis is complete, other practical methods can be employed to improve the purity. Solid products can be recrystallised to remove impurities, while oils and solids containing volatile contaminants can be placed under a vacuum to remove residual solvent and other volatile species.

The purpose of a **recrystallisation** is to improve the purity of the product that has been synthesised and will thereby generate a more meaningful analysis/characterisation (spectroscopy, melting point, etc.) or, in an industrial setting, a product of suitable quality to manufacture. The general approach to recrystallising a previously unknown material is to select an appropriate solvent that will dissolve both the product and the small amount of contaminant, with heating, and result in the product compound being precipitated out, on cooling or by introduction of the anti-solvent. Insoluble impurities are separated by hot filtration.

The solute must be dissolved in the minimum amount of solvent and the amount of solute that can be dissolved increases with temperature, hence heating. The solute is heated in the solvent close to its boiling point, at which the maximum amount of solute can be dissolved, then upon cooling the solution becomes fully saturated at lower concentrations of solute, which, therefore, cannot be held in solution, so precipitates out as purified crystals. The process of crystallisation requires a 'seeding' site to initiate the growth of the crystals, which nucleates from this site and can be achieved by simply scratching the surface of the glass wear. These crystals can later be collected by filtration and the impurities remain in solution and are discarded in the filtrate.

Selecting the right solvent is crucial for a recrystallisation. The product must be insoluble at room temperature, but soluble with heating, while the impurities must either be insoluble and removed by hot filtration, or soluble, remaining in solution at room temperature. Selecting a suitable solvent may involve some trial and error, but principally a polar solute will be dissolved in a polar solvent (similarly a non-polar solute will dissolve in non-polar solvents). However, the polarity of the product and solvent should not be too closely matched as the product must be insoluble at room temperature. For mixed solvent recrystallisations, the anti-solvent is selected so that the product is insoluble within it, i.e. opposite polarity, and consequently will precipitate out of solution when an excess of anti-solvent is added.

In the event that the initial method fails to yield any isolatable material, a number of techniques can be applied to stimulate crystal growth. The solution can be cooled in an ice bath to reduce the product's solubility, but slow grown crystals at room temperature are purer. Alternatively, scratching the glass wear, producing fine shards to act as nuclei for crystal growth may help. If there is a large excess of solvent, crystals will not grow, so some of the solvent will have to be evaporated by spreading on a lager surface (larger beaker/watch glass) or, if necessary, by heating/rotary evaporation. If none of the above work, then it may be necessary to revise the choice of solvent.

TUTORIAL QUESTIONS

1. Explain the terms 'homologous series' and 'isomers'.
2. Show the structural and displayed formula for all the possible isomers with the molecular formula C_6H_{14}. Name each of these.
3. Write an equation to show the thermal cracking of $C_{14}H_{30}$ and write a second equation showing the complete combustion of one of the products.
4. Two unknown compounds, A and B, have the same molecular formula, $C_3H_6O_2$. IR spectroscopy was used to identify possible structures. Compound A produced peaks at 1720 cm^{-1} and 1030 cm^{-1}. Compound B had a peak at 1700 cm^{-1} and a broad absorbance between 2600 and 3300 cm^{-1}. Identify possible structure for these compounds.
5. What peaks would you expect to see in the mass spectrum of butanone?
6. The NMR spectrum of compound X (C_2H_6O) has a singlet at δ 4.5, a quartet at δ 3.6 and a triplet at δ 1.2. Identify this compound.

2 Reactions of Organic Molecules

ABSTRACT

Organic functional group chemistry typically follows characteristic reaction mechanisms. Knowledge of how different functional groups react under different reaction conditions is essential for chemical synthesis. Halogenoalkanes will react via nucleophilic substation or elimination mechanisms, depending on the type of alkyl halide and reaction conditions. Alkenes can be prepared from halogenoalkanes and will typically react with electrophile via and electrophilic addition reaction, which can be used to introduce new functional groups to the carbon chain. Carbonyls are a versatile functional group that may undergo various kinds of reactions, namely nucleophilic addition, alpha-substitution reactions and acyl substitution. Aromatic compounds are also very common in organic synthesis and often comprise the scaffold for useful compounds. Due to the stability of benzene versus alkenes, the ring will react via electrophilic aromatic substitution reactions. This is a useful way of introducing new groups to the benzene ring. Organic chemistry is also paramount for the polymer industry, where alkene monomers can be combined to make addition polymers, or monomers may react with neighbours to form condensation polymers, offering a wide variety of materials that can be made.

Reactions of organic molecules are for the most part governed by the presence of heteroatoms. They have the ability to disturb the electron density within the local area of the hydrocarbon skeleton and therefore create a reactive centre. The positioning and nature of the bonding of heteroatoms in organic molecules are identified as functional groups, which will undergo characteristic reactions.

Understanding electrons is essential to chemistry. In a reaction, chemical bonds must be broken, and this may be a heterolytic cleavage, where two electrons in the bond move to one species to form ions, or a homolytic cleavage, where the pair of electrons is shared to produce free radicals. In organic chemistry, the movement of electrons is shown with curly arrows to produce organic reaction mechanisms, which will feature later in the text. Since reactivity is the movement of electrons to break weak bonds and make new, stronger bonds, it is possible to predict how an organic reaction mechanism will proceed. For two reacting molecules, identify where the electrons are coming from. This molecule is termed the nucleophile: a negatively charged ion, or neutral molecule with a lone pair of electrons which are donated to form a covalent bond. The electrons are received by the electron-deficient molecule called an electrophile. Whether a given molecule will react as a nucleophile or an electrophile depends on the functional groups that are present.

Organic reactions can be classified as either acid–base reactions or redox reactions. The transfer of a hydrogen ion (a proton) identifies an acid–base reaction, while a change in functional group from reactants to products shows reduction–oxidation

DOI: 10.1201/9781003516835-2

reactions. Processes that involve homolytic bond cleavage are called radical reactions. Processes that involve heterolytic bond breaking are called polar reactions. Polar reactions are the most common type of reaction of organic molecules and involve reactions between polar molecules and ions. There are three main classes of polar reactions. An addition reaction involves the combining of two molecules to yield a single product, while in an elimination reaction, one reactant molecule is converted into two product molecules. In a substitution reaction, one functional group on the molecule is replaced by another.

2.1 HALOGENOALKANES

Halogenoalkanes have an alkane skeleton, with one or more halogen atoms attached. Because the halogens are very electronegative, the $C^{\delta+} - X^{\delta-}$ bond is polar. This influences the physicochemical properties of halogenoalkanes. The solubility of halogenoalkanes is determined by the polarity of the C–X bond and the parent chain length. The C–X bond is not polar enough for them to dissolve in water, but they will dissolve in hydrocarbons, making them useful in dry-cleaning fluids. The boiling point of halogenoalkanes depends on the carbon chain length, as greater numbers of electrons lead to greater Van der Walls forces, so longer chains have greater boiling points. The boiling points of halogenoalkanes relative to alkanes of the same carbon chain length are greater, as the halogen atom provides more electron density and causes the molecule to have a greater M_r so Van der Walls forces are greater, and the electronegativity of the halogen produces a dipole–dipole interaction.

The presence of a halogen atom has a significant effect on the reactivity of these molecules relative to alkanes. The reactivity depends on (1) the bond polarity and (2) bond enthalpy. The halogen atom is electronegative so withdraws electrons from the carbon atom, causing it to have a partial positive charge. This makes it susceptible to negatively charged reactive species, called nucleophiles. Additionally, the C–X bond enthalpy decreases down the group, so larger halogen atoms, such as iodide, have a weaker bond to carbon because the shared electron density is further from the halogen nucleus. This makes the bond easier to break, so will influence the mechanism of reactivity.

Due to the bond polarity of the C–X bond, where the electronegative halogen atom withdraws electrons from the carbon, causing it to have a partial positive charge, halogenoalkanes will react with species that are able to donate a pair of electrons. These are termed nucleophiles.

A **nucleophile** is a reagent with a lone pair of electrons with which it forms a covalent bond when it attacks a $C^{\delta+}$ carbon atom.

Examples of nucleophiles include bases such as sodium hydroxide, ammonia and the cyanide ion. **Nucleophilic substitution** reactions are important in organic synthesis, where the introduction of new functional groups makes intermediate chemicals. Alcohols can be made from halogenoalkanes using a weak base, such as sodium hydroxide, to introduce the hydroxyl group, as shown in the reaction mechanism in Figure 2.1. A polar solvent is used, such as an aqueous mixture of ethanol at room temperature.

Organic reaction mechanisms are used to demonstrate the movement of pairs of electrons during a reaction, shown by double-headed curly arrows. In the reaction mechanism for nucleophilic substitution of the halide for the hydroxyl group, the hydroxyl group attacks the $C^{\delta+}$ carbon atom of the C–X bond with its lone pair of

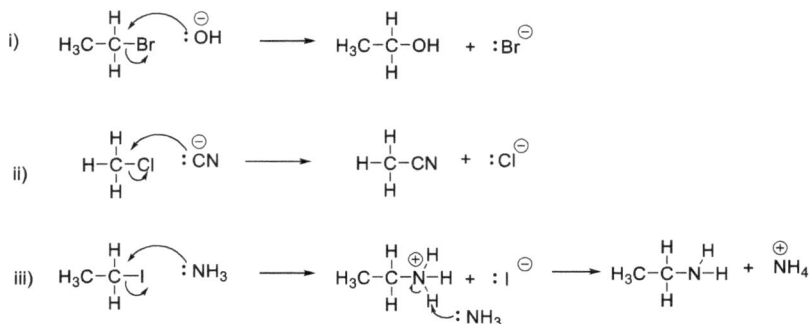

FIGURE 2.1 Examples of reaction mechanisms for nucleophilic substitution: (i) with the hydroxyl ion, (ii) with the cyanide ion and (iii) with ammonia.

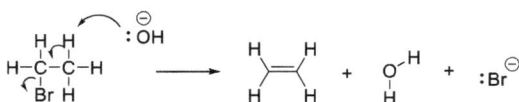

FIGURE 2.2 Elimination reaction mechanism showing the formation of an alkene from a halogenoalkane, when the hydroxide ion acts as a base.

electrons to form a covalent bond. Cleavage of the C–X bond occurs simultaneously, as the halide is displaced. This is known as an S_N1 reaction. This process is not always concerted; during an S_N2 reaction, C–X bond cleavage occurs first.

Certain rules can be applied to help propose a mechanism for an organic reaction. Electrons belonging to a nucleophile will always attack a region of positive charge, in this case, the carbon bonded to the halogen. These electrons will ultimately end up going to the most electronegative atom. By applying these rules, organic reactions can be readily predicted and understood.

Under different conditions, the hydroxide ion will act as a base, rather than as a nucleophile. This leads to a different reaction mechanism, called **elimination**. Pure ethanol is used as a solvent to dissolve a strong base, such as KOH, and the reaction mixture is heated. The reaction mechanism is illustrated in Figure 2.2, showing the formation of an alkene from the halogenoalkane. The hydroxide ion deprotonates the α-carbon causing the shared electrons in the C–H bond to form a C=C double bond, as the halide ion is cleaved, which is referred to as the leaving group.

Alkyl halides become more reactive up the group, with more electronegative halogen atoms withdrawing the electron density from the carbon atom to which they are bonded, thus polarising the bond and making it susceptible to attack from nucleophiles. The precise reaction conditions will determine the reaction mechanism that it undertakes and this in turn will govern the choice of reactants in the reaction mixture used during synthesis.

Whether the reaction is a substitution or an elimination depends on the conditions of the reaction and the type of halogenoalkane used. (1) Substitution is favoured using hydroxide ions in aqueous solution at room temperature, using primary halogenoalkanes. (2) Elimination is favoured using a strong base to provide hydroxide ions in an ethanol solvent, heated, using tertiary halogenoalkanes.

Substitution versus elimination reactions: which process occurs depends on the type of alkyl halide being used and the reaction conditions.

- Primary haloalkanes favour E2/S_N2 reactions. S_N2 is favoured with small strong nucleophiles, e.g. anions OH$^-$, CN$^-$. E2 reactions require a strong base, with heating.
- Tertiary haloalkanes favour E1/S_N1 reactions. S_N1 is favoured by small nucleophiles, while E1 prefers a large base, with heating.
- Secondary haloalkanes will react via E1/S_N1 in polar protic solvents. Reaction via E2/S_N2 prefers polar aprotic solvents, with high concentrations of strong nucleophile/base.

2.1.1 MECHANISMS OF NUCLEOPHILIC SUBSTITUTION REACTIONS

The notation S_N2 signifies that the single-step mechanism is bimolecular, which implies that both the haloalkane and the nucleophile are involved in forming the transition state during the rate-determining step; hence rate depends on the concentration of both these species: k[R–X][Nu]. Rear-end attack by the nucleophile forms a new bond, while simultaneous cleavage of the C–X bond occurs via a transition state. For a chiral carbon, this results in an inversion of configuration. This mechanism is disfavoured by tertiary haloalkanes because substituents sterically hinder the rear-end approach of the nucleophile and +I effects push electron density towards the C–X, making it less electrophilic.

The notation S_N1 signifies the two-step unimolecular reaction, where cleavage of the C–X bond produces a carbocation intermediate, which is readily attacked by a nucleophile. Hence rate depends only on the concentration of the haloalkane: k[R–X]. The planar carbocation can equally be attacked from above and below, so a racemic mixture is produced. This mechanism is favoured by tertiary haloalkanes because the +I effects of the alkyl groups stabilise the carbocation, along with release of steric strain. Primary haloalkanes will only undergo S_N1 if the intermediate is stabilised mesomerically; sp^2 hybridised alkyl halides will not undergo any substitution. The reaction profiles for each mechanism are shown in Figure 2.3.

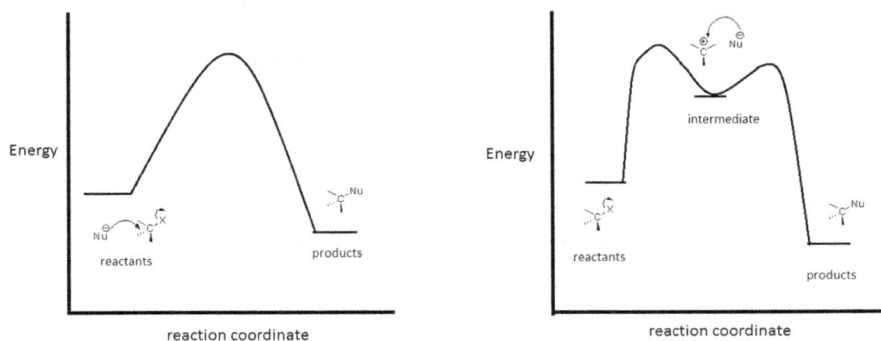

FIGURE 2.3 Reaction profiles for S_N2 mechanism (left) and S_N1 mechanism (right). The comparison of reaction profiles shows clearly the involvement of an intermediate carbocation during the S_N1 mechanism.

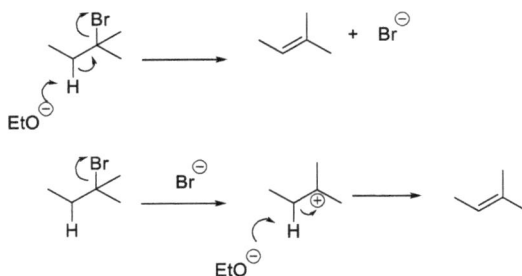

FIGURE 2.4 Elimination mechanisms E2 above, E1 below. These have analogous reaction profiles to those in Figure 2.3.

2.1.2 MECHANISMS OF ELIMINATION REACTIONS

Bimolecular eliminations, E2, involve both the haloalkane and the base in the transition state, k[R–X][B], where the base removes a hydrogen from the alpha carbon, with simultaneous cleavage of the C–X bond to produce an alkene. The reaction is stereoselective, favouring the less strained E-isomer, and regioselective to produce the most substituted alkene in a Saytzeff elimination. However, if a large base is used, it has difficulty approaching the hydrogen on the secondary carbon, so removes the more accessible hydrogen on the primary carbon, known as a Hofmann elimination. Unimolecular elimination, E1, involves formation of a carbocation intermediate, so rate is only dependent on concentration of haloalkane: k[R–X]. The C–X bond cleavage occurs most readily when there are alkyl substituents to stabilise the carbocation, hence the mechanisms favour tertiary haloalkanes. These mechanisms are shown in Figure 2.4.

2.1.3 FREE RADICAL SUBSTITUTION

Halogenoalkanes can be produced from reaction between a halogen and an alkane via a mechanism called **free radical substitution**. Ultraviolet radiation provides the activation energy for the heterolytic cleavage of the halogen, producing a pair of free radicals, which go on to react with the alkane in a chain reaction. This mechanism proceeds in three major steps, outlined in the reaction scheme below. The chlorine free radical may react with chloromethane to produce dichloromethane, and so forth. The chlorine free radical has an equal probability of attacking anywhere on a linear alkane chain, so a mixture of isomers will be produced. Consequently, the chain reaction is not very useful for synthesis of long chain halogenoalkanes.

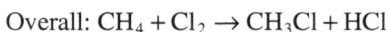

$$\text{Overall: } CH_4 + Cl_2 \rightarrow CH_3Cl + HCl$$

1. **Initiation**: a quantum of UV provides the energy to break the Cl–Cl bond to create two free radicals, as an electron goes to each Cl atom.

$$Cl_2 \rightarrow 2Cl \bullet$$

2. **Propagation**: this takes place in two steps: (i) a chlorine free radical removes a hydrogen atom to produce an alkane free radical; (ii) the alkane free radical then reacts with a chlorine molecule to produce the halogenoalkane and another Cl radical, which will go on to repeat the reaction.

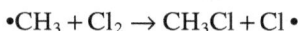

$$Cl \bullet + CH_4 \rightarrow HCl + \bullet CH_3$$

$$\bullet CH_3 + Cl_2 \rightarrow CH_3Cl + Cl \bullet$$

3. **Termination**: two free radicals react to produce a stable compound.

$$Cl \bullet + Cl \bullet \rightarrow Cl_2$$

$$\bullet CH_3 + \bullet CH_3 \rightarrow C_2H_6$$

$$\bullet CH_3 + Cl \bullet \rightarrow CH_3Cl$$

Reactions using free radical substitution mechanism are involved in the production of halogenoalkanes from a halogen and an alkane.

Chlorofluorocarbons (CFCs) are halogenoalkanes where there are no hydrogen atoms present in the molecule; all carbon valences are occupied by bonds to either other carbon atoms or to halogens. These molecules were used in aerosols and refrigerants and as dry-cleaning solvents. In the presence of UV, CFCs break down into free radicals. When this process occurs in the atmosphere, the free radicals can react with ozone in a chain reaction, decomposing it to oxygen. This is illustrated in the reaction scheme below. Ozone in the stratosphere is vital for shielding the Earth from harmful UV radiation, hence CFCs have been banned, due to this association with ozone depletion. These chemicals were replaced by less harmful hydrochlorofluorocarbons (HCFCs).

$$Cl \bullet + O_3 \rightarrow ClO \bullet + O_2$$

$$ClO \bullet + O_3 \rightarrow 2O_2 + Cl \bullet$$

Overall: $2O_3 \rightarrow 3O_2$

Halogenoalkanes can be produced from free radical substitution reactions. However, this produces a mixture of products, so is not the ideal method for organic synthesis. The reactions of halogenoalkanes can be useful in synthesis. For example, the halide can be replaced in a nucleophilic substitution reaction to introduce a new functional group. Also, the elimination reactions of halogenoalkanes are a useful way to make alkenes. These reactions will depend on the conditions of the reaction mixture, so it is important to make a note of these when designing chemical syntheses.

2.2 ALKENES

Carbon atoms can form multiple bonds to other carbon atoms, producing a region of high electron density. This localised region of negative charge is attractive to positively charged reactive species called electrophiles and these reactions can be utilised for the addition of new functional groups to a carbon chain structure during synthesis.

Alkenes are unsaturated hydrocarbons. That is to say, they contain C=C double bonds. Because there is no rotation around a double bond, due to the existence of a π -orbital, the C=C bond has a planar structure, and geometrical isomers can exist. The position of the double bond can of course be in different positions along the hydrocarbon chain, so **position isomers** can exist. For example, but-1-ene and but-2-ene.

Geometrical isomerism is where the molecules have the same structural formula, but the atoms are arranged differently in space.

In the case of but-2-ene, each side of the carbon chain can be on opposite sides of the double bond. This is the E (trans) isomer. If the carbon chain is on the same side of the double bond, then it is the Z (cis) isomer. These isomers are illustrated in Figure 2.5.

Because of the enhanced electron density that is present in a C=C double bond, alkenes are more reactive than alkanes. Typically, they will react with positively charged species that are willing to accept electrons, known as electrophiles. This reaction involves the addition of the electrophile to the carbon chain, hence it is known as an **electrophilic addition** reaction.

Electrophiles are positively charged reagents that are attracted to the electron density of the C=C double bond and accept a pair of electrons.

For example, alkenes will react with hydrogen halides. The reaction mechanism for this is shown in Figure 2.6. A pair of electrons from the C=C double bond attack the hydrogen of the hydrogen halide, as it has a partial positive charge due to its electrons being drawn towards the electronegative halogen. This results in cleavage of the H–X bond. This generates a **carbocation** intermediate. The halide leaving group then attacks the carbocation, forming a covalent bond.

With asymmetrical alkenes, because the halide can potentially attach on either side of the C=C double bond, several products can be made. However, the product mixture consists almost entirely of the halogenoalkane made from the most stable carbocation intermediate. The order of stability is tertiary > secondary > primary

E-but-2-ene   Z-but-2-ene

FIGURE 2.5   The geometric isomers of but-2-ene.

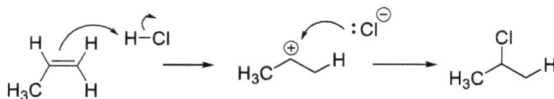

FIGURE 2.6   Reaction mechanism for the electrophilic addition reaction of an alkene with a hydrogen halide.

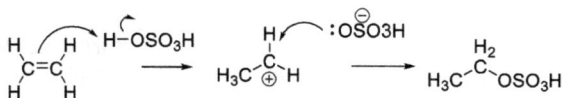

FIGURE 2.7 Mechanism for the reaction of ethane with concentrated sulphuric acid to make ethanol. Water can then be added to reform the sulphuric acid, which serves as a catalyst for the reaction.

due to the **positive induction effect (+I)**; the alkyl groups have a tendency to release electron density and stabilise the charge on the positive carbocation. Hence the product that is formed from the reaction in Figure 2.7 is 2-chloropropane, not 1-chloropropane. This **reaction selectivity** means that this is a good method of synthesising halogenoalkanes.

Carbon atoms can form multiple bonds to other carbon atoms, producing a region of high electron density. This localised region of negative charge is attractive to positively charged reactive species called electrophiles and these reactions can be utilised for the addition of new functional groups to a carbon chain structure during synthesis.

2.2.1 REACTIONS OF ALKENES

The electron density of the double bond makes alkenes susceptible to attack by electrophiles. They will readily undergo a range of addition reactions that can be used in synthesis to introduce new functional groups to a target molecule. A selection of the reactions undergone by alkenes are summarised in the reaction schemes below, which details the reaction mechanisms of various kinds of addition reactions.

1. Electrophilic addition with halogens
   i. Via halonium ion.
      The addition of bromine occurs exclusively as anti-additions, due to the intermediary bromonium ion. Br⁻ acts as a nucleophile and opens the bromonium ion in an S_N2 reaction: inversion of the stereocentre means anti-addition to produce a racemate of trans isomers.

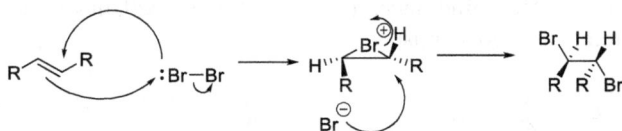

   ii. Halohydrin formation
      If bromine water is used, H_2O can intercept the bromonium ion to form a bromohydrin. This is regioselective, as H_2O nucleophile attacks the most substituted carbon.

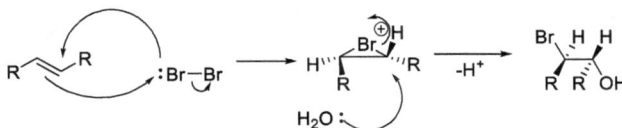

iii. Addition of hydrogen halides

For asymmetric alkenes, regioselective to form the Markovnikov product, the halogen attaches to the most substituted carbon to stabilise carbocation.

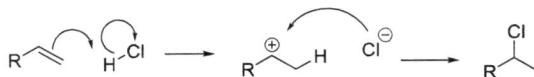

2. Radical addition of HBr

When a peroxide is added, the reaction is selective for the anti-Markovnikov product; as Br adds to the less substituted, more stable, carbon-centred radical.

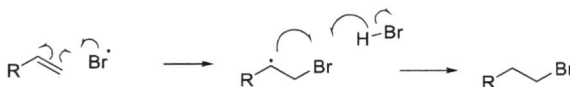

3. Hydration of alkenes
   i. Oxymercuration

Mild reaction conditions with $Hg(OAc)_2$ forming mercurinium ion, which is opened by water to form most substituted product.

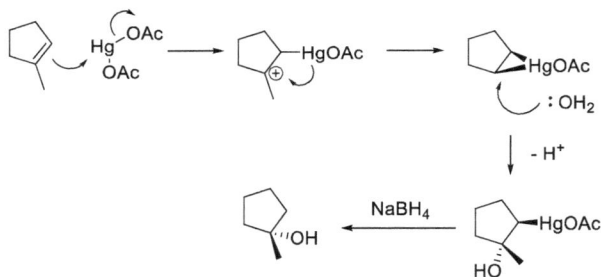

   ii. Hydroboration

Anti-Markovnikov product as BH_3 adds to least substituted carbon due to steric hindrance, in a syn addition. Oxidation of boron to alcohol retains stereochemistry.

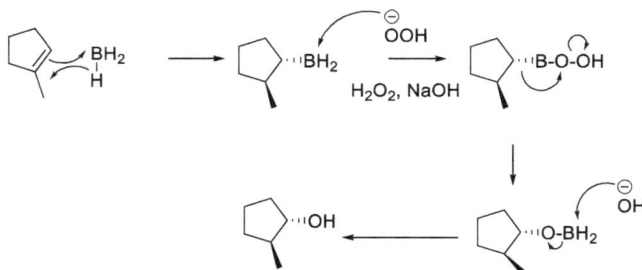

4. Dihydroxylation of alkenes
    i. Epoxidation then ring opening
        Alkenes are readily converted to epoxides by peroxycarboxylic
        acids. S$_N$2 ring opening by water then gives a trans diol.

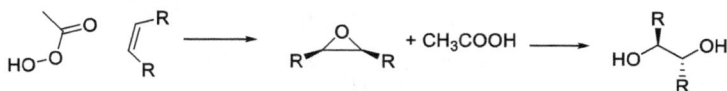

    ii. Direct dihydroxylation
        Concerted mechanism for addition of OsO$_4$ across alkene, producing
        a cis diol.

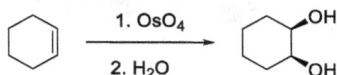

5. Diels–Alder cycloaddition
    If a substituted dieneophile is used, its stereochemistry is transferred into
    the cyclic product.

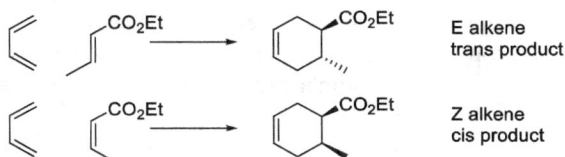

E alkene
trans product

Z alkene
cis product

6. Cross-coupling reactions
    Coupling of a vinyl halide with either a vinyl stanine (Stille) or a vinyl
    boronic acid (Suzuki) to form dienes.

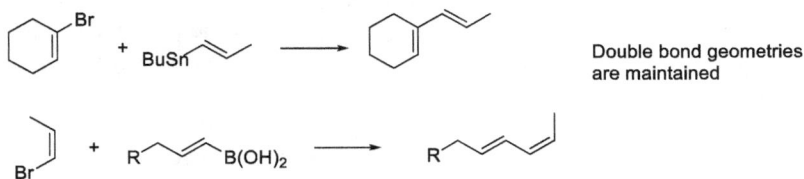

Double bond geometries
are maintained

7. Alkene metathesis
    Combines two alkenes, general reaction:

8. Ozonolysis
    Ozone may be used to convert a C=C bond into two C=O bonds.

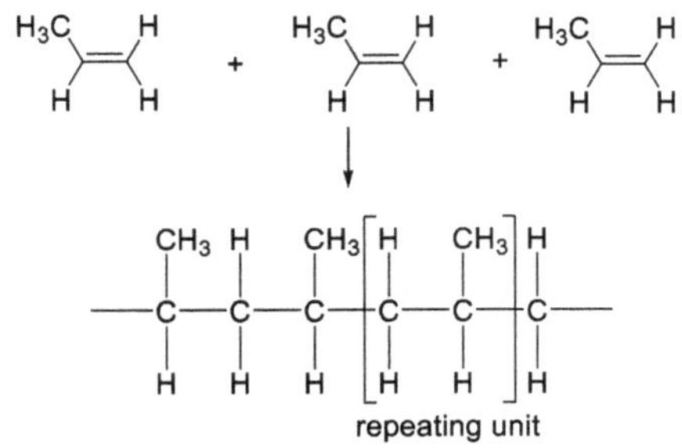

FIGURE 2.8   Reaction scheme showing the addition polymerisation of propene monomers to form the polymer polypropylene, with the repeating unit highlighted in the structure.

Alkenes can also operate as **monomers** that bond together to make **polymers**. This is known as **addition polymerisation**. Polymers are among some of the most useful materials in modern life, so understanding their chemistry is important. The plastics industry manufactures polymers that are used in a variety of applications including packaging, construction, electronics, aerospace and transportation. For example, polypropylene, Figure 2.8, is a hard plastic that is resistant to heat fatigue, making it useful for consumer grade appliances, such as kettles, and piping systems.

However, resistance to degradation means that plastic waste has become a problem with regard to the environment. The materials will persist in landfill and combustion produces obnoxious fumes, which pollute the atmosphere. To mitigate these problems, recycling initiatives need to be implemented. Mechanical recycling involves grinding down separated plastic materials into pellets, which can be melted down and remoulded. However, this process can degrade the properties of the material. Feedstock recycling involves heating the plastics so that the polymer bonds break, and the resulting monomers can be used to make new plastics, but this process is more expensive.

Alkenes are characterised by the presence of a C=C double bond. There is no rotation possible at the double bond, which means that geometric isomers can arise. This can have implications for the chemical reactivity of alkenes. Alkenes will undergo electrophilic addition reactions at the electron-rich double bond, as the electrons attack an electrophilic species to selectively produce a product from the most stable carbocation intermediate. The reactions of alkenes are important for making polymers. Alkenes are also incredibly important in the plastics industry, where they serve as monomers used to manufacture polymers, with application in a variety of products in different industrial sectors.

2.2.2   PREPARATION OF ALKENES

Alkenes can be prepared in different ways including the reduction of alkynes and from elimination reactions of halogenoalkanes and alcohols. Examples of the preparation of alkenes are given in the reaction below.

1. From partial reduction of alkynes
   i   Using a Lindlar catalyst.

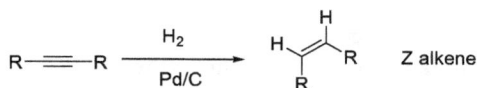

ii   Using sodium dissolved in liquid ammonia.

2. From halogenoalkanes and alcohols
   Haloalkanes and alcohols undergo elimination reactions. Alcohols require a catalyst to produce a good leaving group in a dehydration reaction.

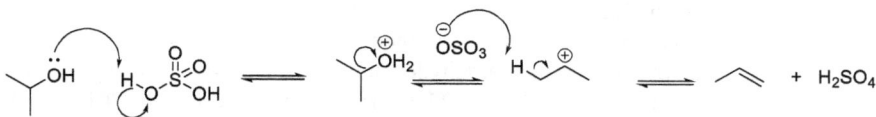

3. From aldehydes and ketone (Wittig reaction)
   The carbonyl is replaced with a phosphonium ylide, with total control over regioselectivity. If R groups are ylide stabilising, E is the major product alkene.

2.3   ALCOHOLS

Alcohols are characterised by the presence of the hydroxyl, OH, group attached to the carbon chain. This means that hydrogen bonding can occur, therefore alcohols have a higher boiling point than alkanes of a similar molecular weight, and smaller chain alcohols are soluble in water. Two key types of reactions undergone by alcohols are **oxidation** and **elimination** reactions.

Oxidation reactions involve an increase in the functional group level of the alcohol, whereby there are more bonds formed between the carbon atom and oxygen atoms. This usually results in the formation of the C=O double bond, known as the carbonyl group. Primary alcohols are oxidised to **aldehydes**, which can be further oxidised to **carboxylic acids**. Secondary alcohols are oxidised to **ketones**. This is done using an appropriate oxidising agent [O], such as acidified potassium dichromate(vi). These reactions are shown in the following worked example.

Worked example: Give equations for the oxidation of ethanol to ethanal and ethanoic acid in separate reaction conditions, and the oxidation of propanone.

Answer: $CH_3CH_2OH + [O] \rightarrow CH_3CHO + H_2O$ *using dilute sulphuric acid,* $K_2Cr_2O_4$ *with gentle heating*

$CH_3CH_2OH + 2[O] \rightarrow CH_3COOH + H_2O$ *using concentrated sulphuric acid, with excess* $K_2Cr_2O_4$ *with gentle heating and reflux*

$CH_3CH(OH)CH_3 + [O] \rightarrow CH_3COCH_3 + H_2O$ *using concentrated sulphuric acid, with excess $K_2Cr_2O_4$ with gentle heating and reflux*

Simple laboratory tests can be performed to distinguish whether the product of the reaction is an aldehyde or a ketone. These tests are summarised in Table 2.1.

Elimination reactions involve the loss of a small molecule from the parent molecule, in this case water, hence the reaction is called a **dehydration** reaction. An example is given in the worked example below.

Worked example: Provide an equation for the dehydration of propanol to produce propene and reaction conditions.

Answer: $CH_3CH_2CH_2OH \rightarrow CH_3CHCH_2 + H_2O$ *using hot concentrated sulphuric acid or by passing the vapours over aluminium oxide heated to 600 K*

Dehydration reactions of longer chain alcohols may produce a mixture of alkenes, showing structural and stereoisomerism. This is demonstrated in Figure 2.9, where the elimination of water can be taken from hydrogen atoms at either side of the hydroxyl group, producing a mixture of alkene products.

Alcohols are characterised by the presence of the OH functional group. They will react via oxidation to produce aldehydes, ketones and carboxylic acids, and by elimination reactions to form alkenes. These reactions are very useful in organic synthesis. Understanding the chemistry of different types of functional groups,

---

TABLE 2.1

Summary of the Tests That Can Be Used in the Laboratory to Distinguish Aldehydes from Ketones

Test	Description	Observations
$H^+/K_2Cr_2O_4$	Further oxidises aldehyde to carboxylic acids, but not ketones	Orange-to-green colour change
Tollens' (Silver mirror)	Gentle oxidation by silver nitrate in aqueous ammonia with warming of aldehyde, but not ketones	Colourless solution to metallic silver deposit
Benedict's	Contains Cu(ii) complex ions, which are reduced on heating with an aldehyde, but not ketones	Blue solution to brick-red precipitate

---

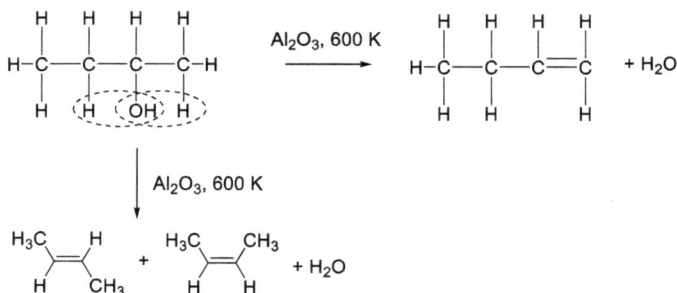

FIGURE 2.9 Reaction scheme for the dehydration of butan-2-ol, showing a mixture of possible products.

their characteristic reactions, the products that are formed from these reactions and the conditions required for the reaction to occur is of the utmost importance for an organic chemist. This knowledge enables chemists to design synthetic strategies to make novel compounds and this skill has application in a multitude of industries, such as pharmaceuticals, agrochemicals and polymer industry.

2.4 CARBONYL COMPOUNDS

Organic chemistry revolves around carbon atoms and the heteroatoms which commonly bond to organic structures. Indeed, it is the heteroatoms that govern the reactivity of organic molecules. These heteroatoms can be incorporated into the carbon skeleton in different ways producing a variety of functional groups, each with distinct reactivity and chemical characteristics. The nomenclature of organic compounds is based on the functional groups that are present. Where there is a double-bonded oxygen to an α-carbon atom, the family of compounds is called the carbonyls. The nomenclature of carbonyls is summarised in Table 2.2.

The carbonyl group involves a carbon atom doubly bonded to an oxygen atom, resulting in a reactive functional group due to the electron density in the double bond being pulled towards the electronegative oxygen atom and the consequential partial positive charge on the alpha carbon atom. The C=O carbonyl group is present in several functional groups, including aldehydes, ketones, carboxylic acids, amines and esters, which undergo characteristic reaction mechanisms that are especially useful in organic synthesis.

TABLE 2.2
Some Common Functional Groups for Advanced Chemistry and Their Associated Nomenclature

Name	Functional Group	Prefix/Suffix
Acyl chloride	$\overset{O}{\underset{R}{\|\|}}\!\!\diagdown\!Cl$	-oyl chloride
Acid anhydride	$R\diagdown\overset{O}{\|\|}\diagup O\diagdown\overset{O}{\|\|}\diagup R$	-oic anhydride
Amine	$R\frown NH_2$	Amino-
Amide	$\overset{O}{\underset{R}{\|\|}}\!\!\diagdown\!NH_2$	Amide
Ester	$\overset{O}{\underset{R}{\|\|}}\!\!\diagdown\!O\!-\!R$	-oate
Carboxylic acid	$\overset{O}{\underset{R}{\|\|}}\!\!\diagdown\!OH$	-oic acid
Nitrile	$R-\!\!\equiv\!\!N$	-nitrile
Aldehyde	$\overset{O}{\underset{R}{\|\|}}\!\!\diagdown\!H$	-al
Ketone	$\overset{O}{\underset{R}{\|\|}}\!\!\diagdown\!R$	-one

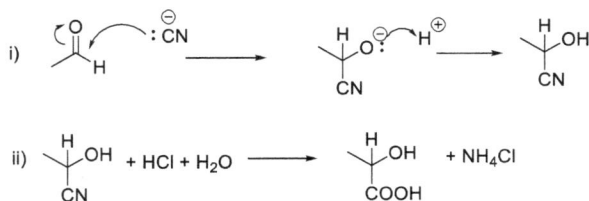

FIGURE 2.10   The nucleophilic addition reaction mechanism for ethanal reacting with the cyanide ion.

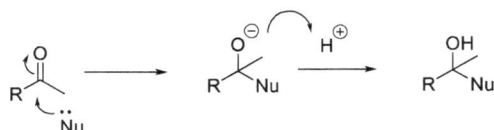

FIGURE 2.11   Generic mechanism for nucleophilic addition to the carbonyl group.

2.4.1   ALDEHYDES AND KETONES: NUCLEOPHILIC ADDITION REACTIONS

Carbonyl compounds, such as aldehydes and ketones, contain the C=O bond. The polarity of this bond gives carbonyls their properties. They have a higher boiling point than alkanes of a similar M_r, short chains are soluble in water and the carbon C=O bond is susceptible to nucleophilic attack. For example, in Figure 2.10 the cyanide ion attacks the $C^{\delta+}$ of ethanal; electrons move from the C=O bond onto the oxygen, giving it a negative charge. The oxygen then uses a lone pair of electrons to form a bond with a proton present in the solvent. This mechanism is called a **nucleophilic addition**.

Aldehydes and ketones react readily by this generic mechanism, shown in Figure 2.11, where Nu can represent a reducing agent, such as $LiAlH_4$ or $NaBH_4$ to make primary and secondary alcohols, respectively; a carbon nucleophile (important for making C–C bonds in synthesis), such as HCN or Grignard's, R–MgX. Equally oxygen nucleophiles can be used, e.g. an alcohol to make a hemiacetal.

Notice that this is a synthesis of an optically active compound. There are four different groups around a central chiral carbon atom. There is no favour for either enantiomer forming, as the CN^- has an equal probability of attacking the flat C=O bond from either side, so a racemate is produced. This reaction is valuable in organic synthesis because it increases the carbon chain length. The scheme in Figure 2.10 shows how the CN functional group can be interconverted to another functional group. This illustrates a synthetic approach to making 2-hydroxypropanoic acid, also known as lactic acid. Interestingly, the lactic acid in milk is the opposite enantiomer to the one present in fatigued muscle. Biological systems have an uncanny ability to produce one enantiomer, whereas in the laboratory this can require intuitive synthetic strategies.

Reduction of aldehydes and ketones also follows the nucleophilic addition mechanism. Several reducing agents, including sodium tetrahydridoborate, which generates the H^- nucleophile, can reduce aldehydes and ketones to alcohols, as shown in Figure 2.12.

$$CH_3CHO + 2[H] \longrightarrow CH_3CH_2OH$$

$$CH_3COCH_3 + 2[H] \longrightarrow CH_3CH(OH)CH_3$$

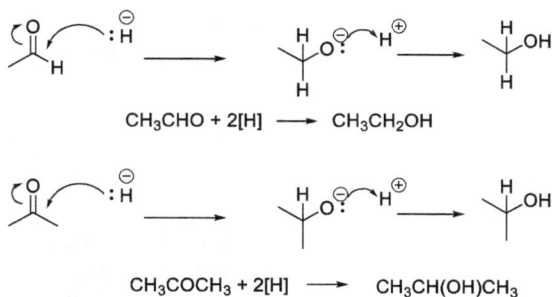

FIGURE 2.12 The reduction of an aldehyde and a ketone via the nucleophilic addition mechanism with the hydride ion.

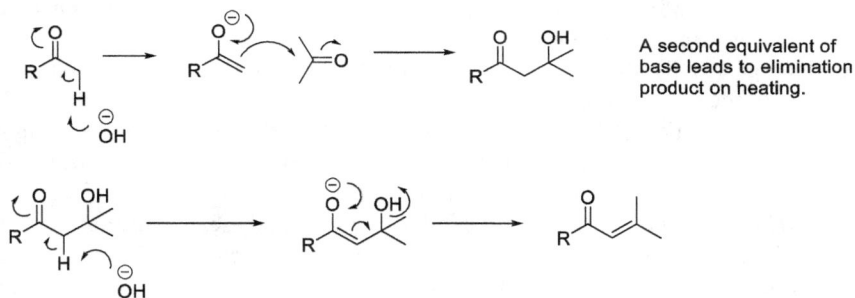

A second equivalent of base leads to elimination product on heating.

FIGURE 2.13 Mechanism for carbonyl alpha-substitution reactions.

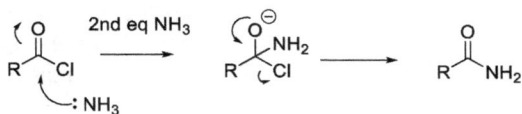

FIGURE 2.14 Mechanism for nucleophilic acyl substitution reactions.

2.4.2 CARBONYL ALPHA-SUBSTITUTION REACTIONS

Removal of the alpha hydrogen by a base produces an enolate, which can act as a nucleophile, as shown in Figure 2.13, and react with electrophiles, for example, the aldol condensation.

2.4.3 NUCLEOPHILIC ACYL SUBSTITUTION

In this reaction, a carboxylic acid derivative is converted into another carbonyl by eliminating a good leaving group, e.g. acyl chlorides, as shown in Figure 2.14.

A range of reactions can be used in organic synthesis, starting from aldehydes and ketones, to produce structures that are medicinally useful. It is important when applying carbonyl chemistry into synthetic routes to consider the regioselectivity and stereoselectivity of these reactions. Carbonyls are a very versatile functional group that will undergo a diversity of reactions. This is on account of the electronegative oxygen atom that polarises the C=O bond, making the $C^{\delta+}$ susceptible to nucleophiles and the αH easily deprotonated with a base. Predominantly, ketones will undergo nucleophilic addition reactions, such as reactions producing a secondary alcohol with hydrides as well as addition of the nucleophile to the carbonyl carbon. A range of nucleophiles can be used, including alcohols, amines, cyanide and Grignard's reagent, which is particularly useful in synthesis for making new C–C bonds.

Stereoselectivity is important for nucleophilic addition reactions. For example, reactions with Grignard's reagent are diastereoselective because of Cram chelation control: the Mg of the Grignard will coordinate with the carbonyl oxygen and any heteroatoms on the adjacent substituents. This is referred to as a conformation fix and ensures selectivity in synthetic routes. The Grignard then attacks the least sterically hindered trajectory to produce one diastereomer. In the absence of an electronegative heteroatom, the conformational fix arises from placing the largest group perpendicular to C=O (Felkin–Anh model).

Reduction with hydrides, e.g., $NaBH_4$ or $LiAlH_4$, produces a racemic mixture of enantiomers because nucleophilic attacks at either side of the C=O are equally likely. Asymmetric reduction of ketones can be achieved through use of the CBS catalyst, whose stereochemistry controls selectivity of the reaction through a cyclic transition state, which provides the conformational fix.

The αH of a ketone can be readily deprotonated with a base. α-Substitution reactions involve replacement of the αH with another group via an enolate. Useful reactions in synthesis include the aldol condensation, where two ketones can be combined and α-alkylation reactions. In each case, a new C2–C3 bond is formed. Diastereoselectivity in α-alkylation can be achieved using oxazolidinones, a chiral auxiliary, which is a temporary control agent to direct the configuration at the new stereo centre. Another important reaction is the conversion of ketones to alkenes using a phosphonium ylide, as this technique offers total control over regiochemistry. More details on the mechanistic features of these synthetic reactions are covered later in the text.

2.4.4 CARBOXYLIC ACIDS

Carboxylic acids will form hydrogen bonds due to the presence of the OH group. Consequently, they are water soluble and have a higher boiling point than alkanes of comparable molecular weight. Dissociation of a proton from the OH group means that they act as weak acids. In an acid–base reaction, the carboxylate ion will form a carboxylate salt. The reaction schemes below demonstrate these kinds of reactions. Included are: reaction schemes for ethanoic acid reacting with (i) a strong alkali, (ii) a carbonate and (iii) a group 2 metal.

$$\text{i. } CH_3COOH + NaOH \rightarrow CH_3COONa + H_2O$$
$$\text{ii. } CH_3COOH + NaHCO_3 \rightarrow CH_3COONa + H_2O + CO_2$$
$$\text{iii. } 2CH_3COOH + 2Mg \rightarrow 2CH_3COOMg + H_2$$

Reaction of a carboxylic acid with an alcohol will produce an ester (**esterification**). The reverse reaction is a hydrolysis. Adding a base to the reaction mixture prevents equilibrium, as a carboxylate salt is formed. This is how soap is made (**saponification**). Therefore, it can be more useful for industrial application to perform the hydrolysis using a dilute alkali, rather than an acid. These reactions are summarised in Figure 2.15. Soaps are made from the hydrolysis of triglycerides (fats and oils), which is illustrated in the reaction scheme in Figure 2.16.

The reactivity of carbonyls is largely predetermined from the nature of the C=O acyl group, which withdraws electrons from the alkyl chain, making the C=O carbon more positive and therefore vulnerable to attack from nucleophiles, as is the case in **acylation reactions**. Acid chlorides will react vigorously with nucleophiles in acylation reactions, as shown in the mechanism in Figure 2.17.

The lone pair of electrons on the oxygen of the alcohol attack the partial positive carbon of the C=O group, pushing the electrons onto the electronegative oxygen. The next step is deprotonation before the electrons relay from the negative oxygen to reform the double bond to carbon, along with cleavage of the Cl leaving group.

Acid anhydrides will also react via this mechanism. These compounds are often favoured in industry over acid chlorides because they do not react as readily with water, are less corrosive, produce safer by-products (a carboxylic acid vs hydrochloric acid) and are less expensive.

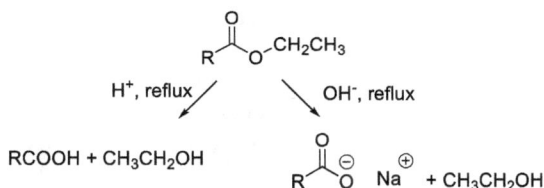

FIGURE 2.15 Reaction schemes showing the hydrolysis of an ester using either a dilute acid or dilute alkali.

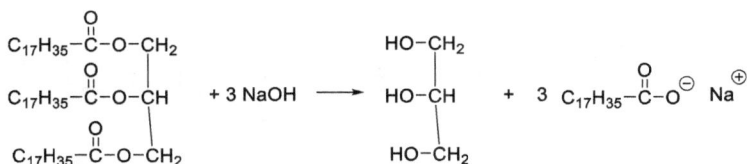

FIGURE 2.16 Reaction scheme for the hydrolysis of a triglyceride during the soap making process.

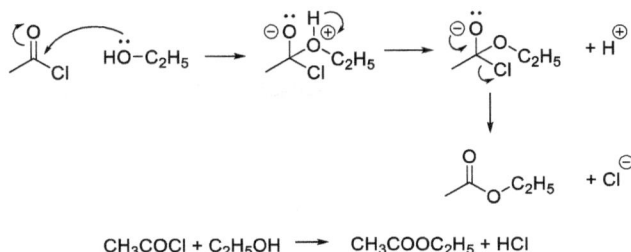

$$CH_3COCl + C_2H_5OH \longrightarrow CH_3COOC_2H_5 + HCl$$

FIGURE 2.17 Reaction mechanism for an acylation reaction between an acid chloride and an alcohol.

Carbonyls are an incredibly important group of compounds in organic chemistry. Their reactions are very valuable in organic synthesis. Nucleophilic addition with aldehydes and ketones can be used to introduce new function groups and use of the cyanide ion will increase the carbon chain length of a product. Carboxylic acids and esters are essential in the manufacture of soap. Acylation reactions are incredibly important for connecting two molecular groups together, towards building a target compound.

2.5   AROMATIC COMPOUNDS

Benzene is the simplest arene. In aromatic compounds, such as benzene, each carbon atom has a delocalised electron in a p-orbital, which moves around the double bonds in a process called resonance. These delocalised electrons collectively form a π system. It is this feature that causes aromatic stability. Evidence for resonance stabilisation of benzene comes from the fact that the C–C bond lengths are intermediate between single and double bonds and that the enthalpy of hydrogenation is lower than expected. Indeed, the enthalpy of hydrogenation would be expected to be three times greater than that of cyclohexene (which is -120 kJ mol^{-1}), assuming normal double bond characteristics, i.e. -360 kJ mol^{-1}, but the measured value is -208 kJ mol^{-1} owing to the stability of the aromatic ring. These processes are summarised in Figure 2.18. Benzene is represented with a Kekulé structure in this text for mechanistic purposes but is also represented as a hexagonal structure containing a circle to represent the delocalisation of the bonding.

Naming aromatic compounds also follows the IUPAC system for nomenclature. Some examples are given in Figure 2.19. These compounds are very useful in

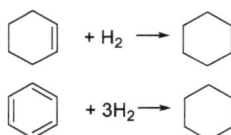

FIGURE 2.18   Hydrogenation of cyclohexene and of benzene.

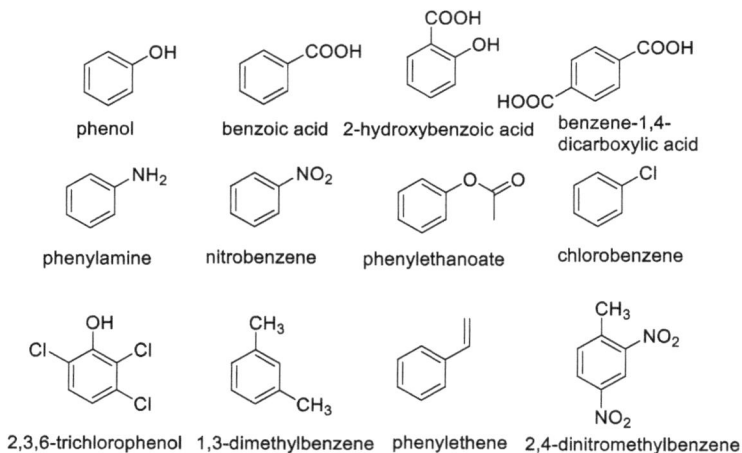

FIGURE 2.19   Examples of aromatic compounds with their IUPAC names assigned.

synthesis, for example, in the pharmaceuticals industry, because the benzene component is often the main scaffold of pharmaceutical molecules, as it holds the organic
substituents in the precise position required for the drug to interact with cell receptors. Many natural products that drug compounds mimic also contain benzene rings.

A characteristic reaction of benzene and its derivatives is **aromatic electrophilic
substitution**. The input of the delocalisation energy is required to break the ring, so
benzene is too stable to react by electrophilic addition. By contrast, the ring is preserved in aromatic electrophilic substitution, as the electrophiles replace a hydrogen
atom. With reference to the examples given in Figures 2.20 and 2.21, electrons from
the π system attack the electrophile to produce a carbocation. A pair of electrons is
taken from the C–H bond to stabilise the charge. A mixture of nitric acid and sulphuric acid is used to produce the **nitronium ion**. Benzene is too stable to react via the
usual acylation mechanism, so in **Friedel–Crafts acylation**, an aluminium chloride
catalyst is used to make the electrophile. This is a very useful method for introducing
organic substituents to a benzene ring.

Understanding the chemistry of benzene is important for an organic chemist.
Being able to use electrophilic aromatic substitution reactions to introduce new substituents to the ring is an essential technique in developing novel compounds, such as
pharmaceuticals. Therefore, a knowledge of the key reactions, such as reaction with
the nitronium ion and Friedel–Crafts acylation, will be a valuable asset. Note that
the Friedel–Crafts acylation serves as a method to introduce new carbon scaffolds
to the molecular structure. This is a frequently used technique in organic synthesis.
Nitration of benzene followed by a functional group interconversion to produce an
amine is valuable in synthesis of pharmaceuticals because it is a way of introducing
hydrogen bonding groups to novel compounds.

Aromatic chemistry revolves around benzene and substituted benzenes. Benzene
is a hexagonal ring of doubly bonded carbon atoms. However, the electrons in the

$$HNO_3 + H_2SO_4 \longrightarrow H_2NO_3^+ + HSO_4^-$$

$$H_2NO_3^+ \longrightarrow NO_2^+ + H_2O$$

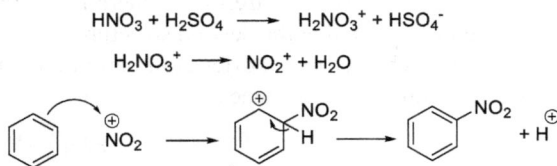

FIGURE 2.20 Mechanism for the nitration of benzene.

$$CH_3COCl + AlCl_3 \longrightarrow CH_3\overset{\oplus}{C}O + AlCl_4^-$$

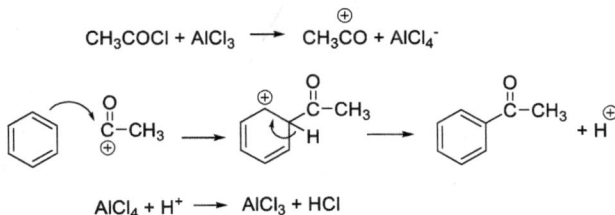

$$AlCl_4 + H^+ \longrightarrow AlCl_3 + HCl$$

FIGURE 2.21 Mechanism for the Friedel–Crafts reaction of benzene.

ring are not fixed to the double bonds and instead are delocalised across the ring due to resonance, forming a π-orbital. Therefore, the reactions of benzene are not characteristic of alkenes and instead benzene undergoes a particular type of reaction mechanism called **electrophilic aromatic substitution**. These are important reactions for synthesis for the creation of substituted benzenes. These types of molecules are incredibly important in medicinal chemistry because they often form the scaffold of pharmaceutical compounds, holding the reactive centres or binding sites firmly in the correct position needed for their therapeutic effect.

2.5.1 ELECTROPHILIC AROMATIC SUBSTITUTION

This is a highly useful reaction in synthesis because benzene will react with various electrophiles and transformation of the substituents can be done. The reaction schemes in Figure 2.22 illustrate the reactions that can be done to add new functionalities to the benzene ring.

Regioselectivity is governed by the other substituents on the ring. Activating groups, which donate electrons into the ring (+I, +M), direct electrophiles to the 2 (ortho) and 4 (para) positions. Deactivating groups, which withdraw electrons from the ring (−I, −M), direct electrophiles to the 3 (meta) position, except halogens, which direct to the 2, 4 positions. Steric factors also control positions of substituents, e.g. favour 4 position.

2.5.2 AROMATIC FUNCTIONAL GROUP CHEMISTRY

The presence of existing functional groups on aromatic ring systems will influence the regioselectivity of subsequent reactions. Different functional groups will direct reactive species towards a particular position on the ring. Therefore, a carefully planned synthetic strategy is required in order to ensure that the added groups are placed correctly in the ring system. Hence careful consideration of the order of reaction processes is important to ensure that the desired structure is produced at the end of the synthesis. The reaction schemes below offer examples of the kinds of reactions that can be performed with substituted benzenes.

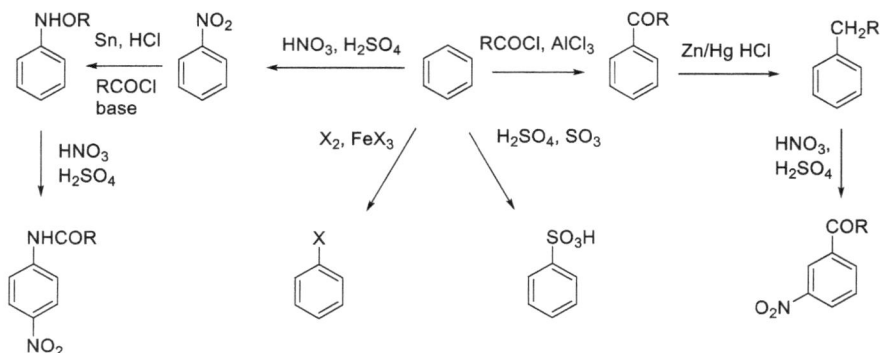

FIGURE 2.22 Synthetic routes for producing substituted benzenes.

1. Nitro group reduction

   Nitrobenzene can be reduced to aniline. This increases the reactivity of the ring and changes the position to which the substituent directs subsequent electrophilic substitutions. Anilines react with acid chlorides to produce an amide (hydrolyses back with the addition of NaOH) which controls the reactivity, hence mono-substitution only is possible.

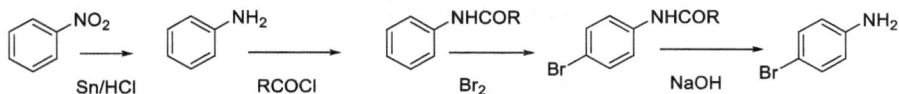

-M, meta director   +M, o,p director   delocalise lp, less reactive   monosubstitution in 4 position

2. Redox reactions

   This is a useful way to change the directing influence of the substituents.

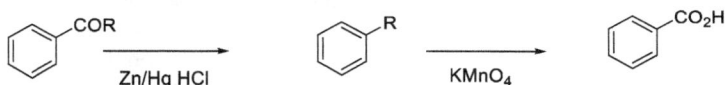

3. Nucleophilic aromatic substitution

   This is in fact an addition–elimination reaction and the −M group and leaving group must be ortho or para to one another in order to resonance stabilise the anion intermediate.

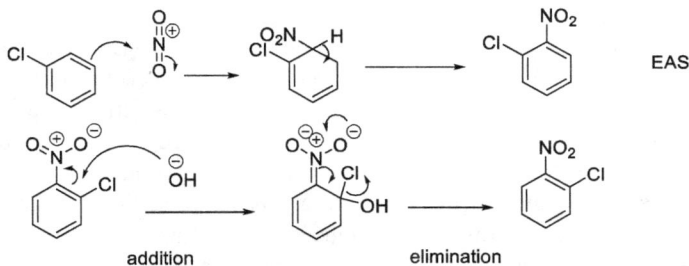

addition   elimination

   Aromatic and heteroaromatic chemistry are of key importance in the pharmaceuticals industry. Several synthetic methods can be utilised to functionalise aromatic ring systems, using electrophiles and/or nucleophiles in a regioselective manner. Aromatic compounds can then be further transformed into more complex structures using carbon–carbon bond forming reactions.

   Benzene commonly reacts with electrophiles during electrophilic substitution reactions, for example, nitration, sulphonation and halogenation, which require a Lewis acid catalyst. A useful example for synthesis would be the Friedel–Crafts reactions (acylation and alkylation) because this forms a new C–C bond.

   Reactions of substituted benzenes depend on the functional groups already present on the ring. Regioselectivity depends on the electronic effects of the substituents. Activating groups are typically +I and +M groups that donate electrons into the ring. Deactivating groups withdraw electrons from the ring (−I and −M groups) and destabilise the ring, making it less reactive. Activating groups direct the reaction to the ortho and para positions, depending on the steric factors of the substituted benzenes, while deactivating groups direct the reaction to the meta position, except halogens, which are 2, 4 directors.

In synthesis, it is therefore necessary to add the substituents in the correct order to achieve the desired regioselectivity. One way to do this is to utilise functional group interconversions, such as nitro group reduction with Sn/HCl to produce aniline, which is a 2, 4 director. Reaction of aniline to produce diazonium salts is used to make dyes and the compounds can be converted into many other functional groups.

2.6   AMINES AND AMINO ACIDS

Amines will form hydrogen bonds, a consequence of the presence of hydrogen atoms being bonded to the electronegative nitrogen atom in the functional group, with tertiary amines being the exception. As a result, small chain amines are soluble in water. The presence of hydrogen bonding means that amines have a higher boiling point than alkanes with a comparable M_r, but a lower boiling point than alcohols with a similar molecular weight, as N is less electronegative than O. The different types of amine: primary, secondary, tertiary are categorised based on the numbers of alkyl chains bonded to the nitrogen atom. Alkyl groups of aliphatic amines donate electrons via the positive inductive effect to the nitrogen, making it a better proton acceptor. Therefore, amines with more alkyl groups act as stronger bases. For example, 2-amino propane, a secondary amine, is a stronger base than ethylamine, a primary amine. Aromatic groups withdraw electrons from the nitrogen into the delocalised π system, so phenylamine is an example of a poor base.

Amines possess a lone pair of electrons on the nitrogen, so will act as nucleophiles. This is illustrated in Figure 2.23. The electrons of the nitrogen attack a partially positive carbon atom in a nucleophilic substitution reaction. The primary amine then goes on to cause further nucleophilic substitution reactions, creating a mixture of primary, secondary and tertiary amines. This is an inefficient way to prepare amines, because fractional distillation is needed to separate these products.

A more efficient means of preparing amines is via reduction of nitriles. In the first step of the synthesis, a nitrile is produced by reacting a halogenoalkane with the cyanide ion in aqueous ethanol at room temperature. This is a nucleophilic substitution reaction. In the second step, the nitrile is then reduced to a primary amine via catalytic hydrogenation (using a nickel catalyst). This yields a pure product of primary amine.

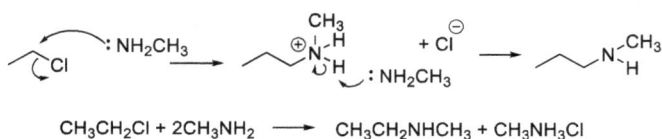

Step 1: $CH_3Br + CN^- \rightarrow CH_3CN + Br^-$
Step 2: $CH_3CN + 2H_2 \rightarrow CH_3CH_2NH_2$

$$CH_3CH_2Cl + 2CH_3NH_2 \longrightarrow CH_3CH_2NHCH_3 + CH_3NH_3Cl$$

FIGURE 2.23 Nucleophilic substitution mechanism for an amine reacting with a halogenoalkane.

The preparation of phenylamine, shown in Figure 2.24, involves a slightly different process. In the first step, nitrobenzene is produced via reaction of nitric acid and concentrated sulphuric acid. In the second step, nitrobenzene is reduced to phenylamine, using tin and hydrochloric acid as reducing agents. These react together to produce the hydrogen that replaces the oxygen atoms.

Amines are economically important molecules. They are used in the polymer industry; nitrogen-containing compounds are highly important in drug design; and quaternary ammonium salts are used in hair and fabric conditioners, as the cationic surfactant.

All naturally occurring amino acids exist as one enantiomer (except glycine, which is not chiral) and are alpha amino acids (amine group is on the carbon next to the COOH group). Because both of these groups, a base and an acid, are present, at a certain pH, known as the **isoelectric point**, the COOH group can donate a proton to the NH_2 to form a **zwitterion**.

The ionic qualities mean that amino acids are solids in an ionic lattice, so have a high melting point and will readily dissolve in water. In acidic conditions, a proton will join the amine group and the amino acids act as a base. In alkaline conditions, they are deprotonated and act as acids. Amino acids will therefore act as both acids and bases and are said to be **amphoteric**.

Amino acids will also polymerise to form peptides. The condensation reaction between two amino acids to form a dipeptide is shown in Figure 2.25. Water is eliminated from the reaction between the amine group of one amino acid and the carboxylate of another to form an amide linkage, or peptide bond, as it is also known. The peptide bonds can be broken by refluxing with 6 M HCl for 24 hours in order to completely hydrolyse all the peptide bonds in the peptide or protein. This is also shown in Figure 2.25.

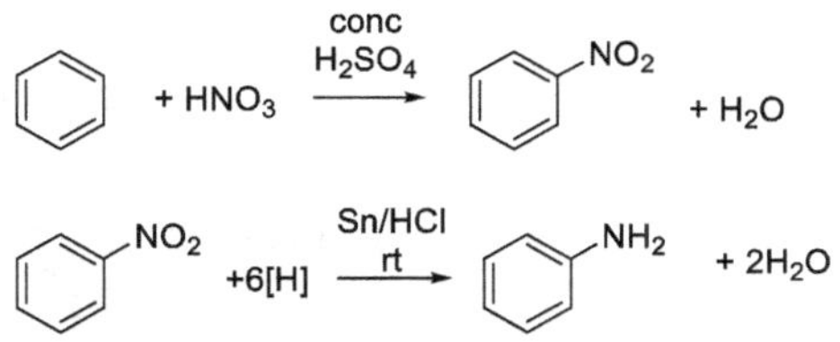

FIGURE 2.24 A synthetic route to produce phenylamine.

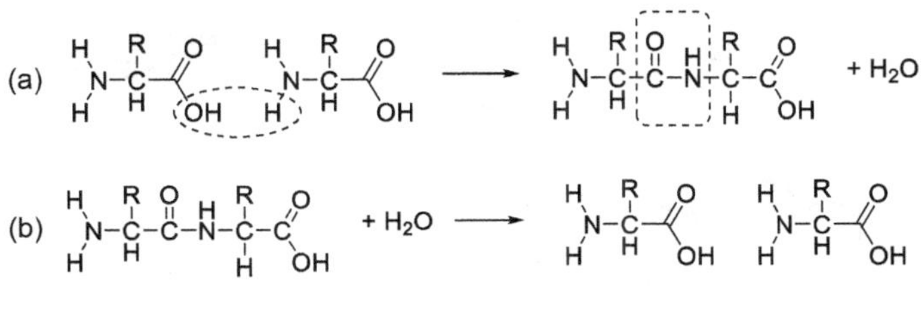

FIGURE 2.25 Condensation reaction between two amino acids to form a dipeptide (a), followed by the hydrolysis of that dipeptide into the constituent amino acids (b).

Proteins consist of long peptide chains, referred to as the primary structure. The precise sequence of the amino acids in the primary structure will determine how and where hydrogen bonds form, as the protein folds into the three-dimensional secondary and tertiary structures. Some proteins may be composed of several peptides and combine in a quaternary structure. The acid–base properties, and the ability of the amine functional group to act as a nucleophile, are important for the reactivity (and therefore function) of proteins, such as enzymes.

2.7　POLYMER CHEMISTRY

Organic molecules can be linked together in long chains to form polymers. This process involves the reaction of functional groups on these molecules to tether the monomers together in the chain. There are different examples of this. Alkenes can be used as monomers, where reaction of the double bond causes it to open up and form bonds to neighbouring molecules. Other examples may involve amine or alcohol or carbonyl functional groups reacting together and eliminating water molecules, as the polymer chains grow.

Addition polymerisation occurs as the double bond of the monomer units opens to enable the bonding to occur between carbon atoms that is necessary for the formation of the polymer backbone. Many modern materials such as polychloroethene (or polyvinylchloride, PVC) and polyphenylethene (polystyrene) are examples of addition polymers (Figure 2.26).

Polyamides are an example of **condensation polymers**, where two functional groups on the monomer units react and expel water to form the backbone of the polymer. Proteins and peptides are polyamides. Synthetic examples include Kevlar and Nylon 6,6 which are fire- and abrasion-resistant materials made from 1,6-diaminohexane and hexane-1,6-dicarboxylic acid, respectively.

Polyesters are also condensation polymers but are made by the formation of an ester link, when reaction between a carboxylic acid and an alcohol takes place between the monomer units. Polyesters are used to make clothing. Terylene is made from benzene-1,4-dicarboxylic acid and ethane-1,2-diol. The structure is given in Figure 2.27.

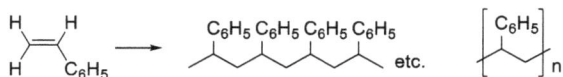

FIGURE 2.26 The chemical structure of polystyrene formed from monomer units of phenylethene, with the repeating unit shown.

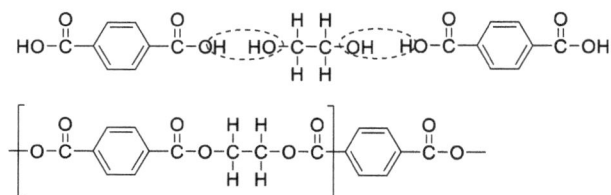

FIGURE 2.27 The chemical structure of terylene, formed by the reaction of benzene-1,4-dicarboxylic acid and ethane-1,2-diol monomers, with the repeating unit shown.

Macromolecules play vital roles in both materials science and the natural world. There are two key ways in which synthetic polymers can be made. Chain growth polymerisation (addition polymerisation) involves addition of an alkene monomer to the end of a growing chain. Step growth polymerisation (condensation polymerisation) involves combining two reactive functional groups on either side of the polymer chain (e.g. esters and amides) while eliminating a small molecule, such as water. Natural polymers, such as proteins, are produced by condensation polymerisation; they are polyamides. However, in the cell, the process is highly complex, mediated by ribosomes and an ensemble of enzymes. Chemists do have methods to synthesise biomolecules in the laboratory, however. For the synthesis of peptides in solution, in order to ensure the correct sequence of amino acids, the N-terminus of the polymer must be protected while the carboxylate will be activated for the coupling reaction. In solid phase peptide synthesis, the C-terminus is attached to an insoluble functionalised polymer bead. The amino acids can then be attached in sequence and finally the complete protein can be detached from the anchoring bead using hydrofluoric acid. Likewise, during the synthesis of polysaccharides (also condensation polymerisation) protecting groups are required to ensure glycosidic bonds are made on the correct positions of the ring.

The structures of polymers are very important because this governs their properties and potential applications. The rigidity of a synthetic polymer depends on the flexibility of the chains. This is greater for single bonds, which are conformationally more mobile, e.g. in polythene. This flexibility can be reduced by the addition of sterically bulky groups, such as benzyl in polystyrene. Another factor to consider is attractive forces between chains. Polyamides, such as nylon and Kevlar, have hydrogen bonds between multiple amide groups, making these polymers very strong, particularly Kevlar, which also contains benzene rings in the polymer backbone. Another method to increase polymer strength is to cross link the chains. By adapting the structures of polymers, chemists can synthesise a variety of macromolecules for a range of applications. Nylon and Kevlar are used in safety clothing, due to their strength, whereas weaker polymers, such as polyester, are used in casual cloths and polythene is used in disposable plastic bags.

Biopolymers have specific roles in the cell. In order to perform their required function, a very precise structure is required. Chemists can make biopolymers synthetically for use in medicine, e.g. insulin, by reproducing the natural structure. A protein's three-dimensional structure depends on the order of amino acids; the unique sequence governs the positions of hydrogen bonds that form and whether the chain folds into an alpha helix or beta sheet. These are the only two structures available due to the nature of the peptide backbone, where bond resonance causes double bond character between the C=N of the amide link, which therefore restricts the number of possible conformations that the chain can adopt. The function of a protein is governed by the tertiary structure which is determined exactly by the particular hydrogen bonding arrangement. Therefore, the functioning of a protein is fundamentally linked to the sequence of amino acids in the polymer chain. DNA contains all the information in the cell for making proteins, so its structure is of the utmost importance. DNA consists of two chains of ribose sugars linked by phosphate groups. The order of the nucleotide monomer bases on the ribose sugars is the instruction manual for making proteins during transcription. Hydrogen bonds between the bases link the two chains together in an alpha helix. Supercoiling of the DNA strands makes it denser, so large amounts of information can be stored in the cell nucleus.

TUTORIAL QUESTIONS

1. Draw the reaction mechanism for bromoethane reacting with sodium hydroxide at room temperature in a polar solvent. Draw the reaction mechanism for bromoethane reacting with potassium hydroxide in ethanol, with heating. Name each of these mechanisms.

2. Write equations outlining the formation of chloromethane from methane and chlorine via free radical substitution. Include all possible termination steps. Comment on the drawbacks of using CFCs.

3. Draw the structures of two possible products of the reaction between 2-methylpent-2-ene and hydrochloric acid. Draw the mechanism for the formation of the major product. Name the mechanism and explain why this is the major product.

4. Write separate equations for the reaction of propanol and propan-2-ol with concentrated sulphuric acid and potassium dichromate(vi). Explain why the ketone is not further oxidised.

5. Draw structures of all the organic products and isomers possible from the dehydration of butan-2-ol. Suggest reagents and conditions for this reaction.

6. Show the alkenes that are structural isomers of the general formula C_4H_8 and draw structures of the E/Z stereoisomers.

7. Show that amino acids, $RCHNH_2COOH$, have a chiral centre by drawing the optical isomers.

8. Draw the structures of ethanenitrile, methylethanoate and methylbutanone.

9. Name the following compounds: CH_3COCl; $HCOOCHO$; CH_3CONH_2; CH_3CH_2COOH.

10. Describe a simple test to distinguish between aldehydes and ketones, suggesting reagents, and including any observations from the test. Explain why ketones do not produce these observations during the test.

11. Write an equation for the reduction of propanone with a generic reducing agent. Name the alcohol that is produced and the type of reaction mechanism. Give an example of one such reducing agent.

12. Write a simple equation to show the formation of propylethanoate from ethanoic acid and an alcohol under acidic conditions.

13. Write a simple equation to show the formation of propylethanoate from ethanoylchloride and an alcohol. Draw the reaction mechanism for this process. Suggest why using ethanoic anhydride instead might be preferred.

14. Explain what is meant by the term 'aromaticity' in terms of the bonding in benzene. Give two pieces of evidence for aromaticity.
15. Write equations for the synthesis of phenylamine from benzene, including the reagents used in each step. Draw the mechanism for the aromatic electrophilic substitution step. Suggest why benzene will less readily react with electrophiles than cyclohexene.
16. Suggest which is the strongest base, 2-aminopropane or phenylamine. Explain your answer.
17. Suggest reaction conditions to hydrolyse a protein into its constituent amino acids.
18. Draw the repeat unit for the polymer Perspex, made from the monomer 2-methylpropenoate. Name this type of polymerisation.

3 Organic Synthesis

ABSTRACT

Organic synthesis requires knowledge of the reaction mechanisms and conditions required for chemical reactions used during synthetic procedures and examines how these reactions can be used in a systematic approach to build a target molecule from readily available starting materials (RASMs). This section covers the principles of advanced synthesis and introduces a variety of chemical strategies that can be used to produce new organic molecules. This includes outlining how retrosynthetic analysis can be implemented to work backwards from a desired target molecule to create a synthetic strategy starting from cost-effective and RASMs. This process involves theoretical bond disconnection to create segments of molecules, referred to as 'synthons'. Synthetic equivalents to these synthons can then be identified that can be reacted together to produce the target molecule. These techniques are critical to the role medicinal chemists play in designing and creating new therapeutic agents.

The synthesis of new organic molecules from a subset of readily available reactants can be achieved by understanding the key organic reactions that are characteristic of different types of organic molecules, often dependent on the presence of particular functional groups. This depends on a number of factors including functional group chemistry, reaction conditions, reaction selectivity and the presence of catalysts, as well as the costs involved in the synthesis. A broad knowledge of organic reactions and the mechanisms of those reactions is therefore invaluable to the organic chemist in order to be able to plan a series of chemical reactions, referred to as a synthetic route, to synthesise a useful target molecule from the readily available reactants.

3.1 RETROSYNTHETIC ANALYSIS

Retrosynthetic analysis is a technique to create a target molecule by retrospectively building the compound from the key organic reactions. A target molecule is first identified, for example, via the use of computer modelling of known drug binding sites, and then the target structure produced in the model can be theoretically broken down into segments. These segments, called synthons, can then be substituted for real chemical structures, which in turn can be synthesised from readily available starting materials (RASMs). Thus, a complete synthetic strategy is made for the target molecule.

3.1.1 SYNTHONS, SYNTHETIC EQUIVALENTS AND FGIS

A target molecule can be broken down into theoretical **synthons**, which are then matched with a **synthetic equivalent**; these may be readily available starting materials or may need to be produced from RASMs (Table 3.1).

To make an appropriate disconnection, and make useful synthons, it may be necessary to first do a **Functional Group Interconversion (FGI)**. This is because not

DOI: 10.1201/9781003516835-3

TABLE 3.1

Illustration of Common Synthons Produced from Theoretical Bond Disconnections of Target Molecules Alongside Their Real Synthetic Equivalents Used During Synthetic Organic Chemistry

Common synthons	Synthetic equivalents

all synthons created by theoretical bond disconnects have useable synthetic equivalents. Therefore, it is often necessary to start from a different functional group and produce and convert it during a later step of the synthesis (Table 3.2).

3.1.2 REACTION SELECTIVITY

Another complication to organic synthesis is reaction selectivity, That is, how to get the target molecule you want. This is not always straightforward because chemical reaction can lead to substituents being placed in alternative position or a mixture of isomers being produced. Therefore, a synthetic procedure must account for reaction selectivity by either choosing reaction processes with the desired selectivity or by implementing techniques that can be used to control reaction selectivity. This is demonstrated in the series of reaction schemes given below.

TABLE 3.2
Common Functional Group Interconversions

Chemoselectivity: a chemoselective reaction involves one functional group reacting in preference to another.

E.g. chemoselective reduction of a ketone (most electrophilic).

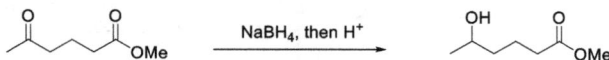

E.g. chemoselective oxidation of an alkene (most electron rich).

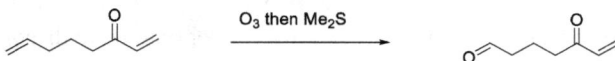

Regioselectivity: a regioselective reaction leads to the formation of one structural isomer.

E.g. regioselective reduction of an enone.

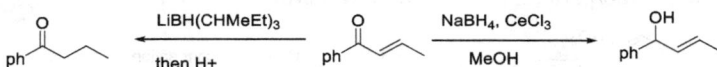

Stereoselectivity: a stereoselective reaction is where one enantiomer, one diastereomer, or one double bond is formed selectively over the other isomer.

Protecting groups: temporary modifications to reactive functional groups can be employed to achieve chemoselectivity.

  i. Alcohols are protected by silyl ethers, which offer regioselective protection; bulky group attacks primary alcohol.

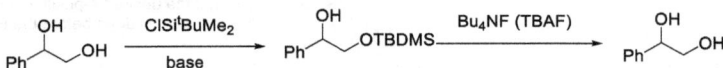

 ii. Carbonyls can be protected by conversion to acetals.

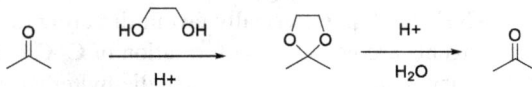

iii. Carboxylic acids are protected with esters, where steric bulk of the group gives chemoselective protection.

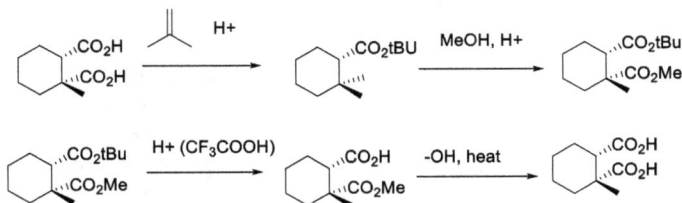

iv. Amines are protected as carbamates, formed by reaction with a carbonyl.

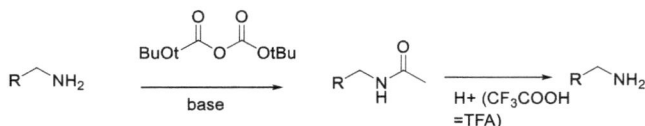

3.1.3 REGIOSELECTIVE SYNTHESIS OF SUBSTITUTED BENZENES

The schemes below show examples of how the appropriate bond disconnections can be made during the retrosynthetic analysis of substituted benzenes, bearing in mind the regioselective nature of their synthesis.

1. To decide the order of disconnections, you must know whether the substituents are ortho/para (2, 4) or meta (3) directing.

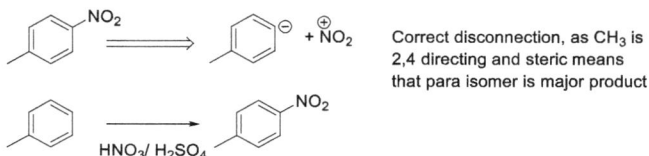

Correct disconnection, as CH_3 is 2,4 directing and steric means that para isomer is major product

2. If an FGI is needed, it may alter the directing effect of the substituent, so decide whether other substituents need adding before or after the FGI.

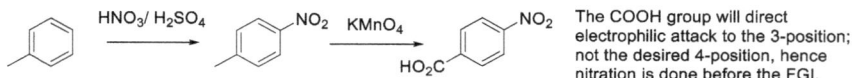

The COOH group will direct electrophilic attack to the 3-position; not the desired 4-position, hence nitration is done before the FGI.

3. OH and OR groups are difficult to add to benzene, so it is often useful not to disconnect them, or use phenol as a starting material.
4. To improve regioselectivity, blocking groups can be introduced, such as converting NH_2 into $NHCOCH_3$, to sterically favour directing to the 4 position.
5. Aryl Grignard reagents are used for the formation of C–C bonds. They are made by reacting aryl halides with Mg in dry diethyl ether and are strong nucleophiles.

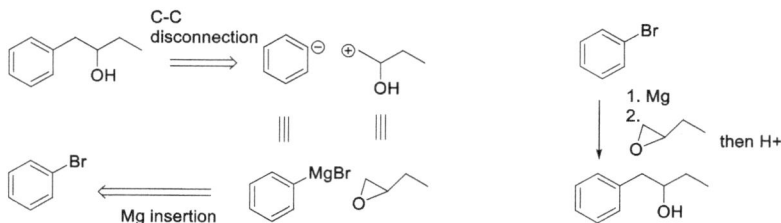

3.1.4 STEREOSELECTIVE PREPARATION OF ALKENES

During the preparation of alkenes, stereoselectivity is an important consideration because control is needed over which geometrical isomer is made. The reaction schemes below illustrate several good choices for preparing alkenes with stereoselectivity.

1. The Wittig reaction is perhaps the most versatile and general method for making C=C bonds.

    Stabilised ylides are prepared by deprotonation of a phosphonium salt, Ph_3PCH_2EWG, with a weak base. They react with aldehydes, not ketones usually, to produce E alkenes.

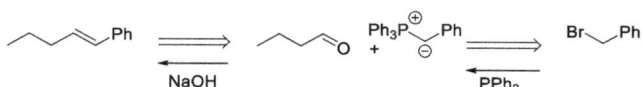

    Unstabilised ylides are prepared by deprotonation of a phosphonium salt, Ph_3PCH_2EDG, with a strong base. They react with aldehydes and ketones to produce Z alkenes.

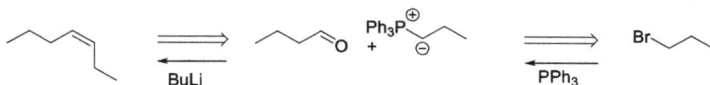

2. Using alkanyl anions, formed by deprotonation of a terminal alkyne with sodamide ($NaNH_2$), then stereoselective reduction (using H_2/Lindlar catalyst to produce Z alkene, or Na/NH_3 to produce an E alkene).

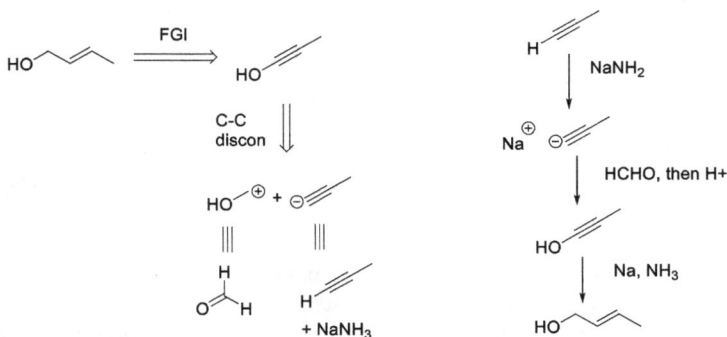

3. **Palladium (0)-catalysed coupling**: the Heck reaction can be used for a disubstituted alkene from a terminal alkene and an aryl halide or vinyl halide (the C atom next to X must be sp^2 hybridised). The R group is introduced at the least hindered end of the double bond to produce the more stable E alkene.

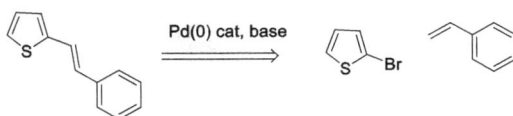

3.1.5 REGIOSELECTIVE REACTIONS OF α, β-UNSATURATED CARBONYLS

These compounds are readily available starting materials, produced via aldol condensations. They are useful in synthesis for making C–C bonds. These reactions
therefore are useful for extending the alkyl chains of target molecules in a regioselective manner, as illustrated in the reaction schemes below.

1. **1, 2-addition (C=O addition)**: Grignard reagents mainly attack directly at
   the C=O.

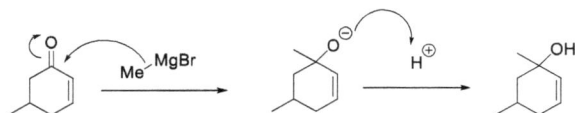

2. **1, 4-addition (conjugate/Michael addition)**: organocopper reagents mainly
   attack at C=C.

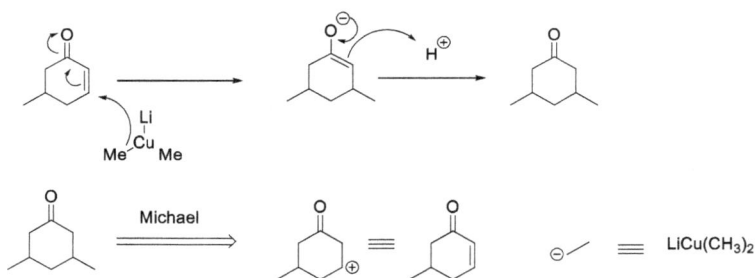

3.1.6 ORGANIC SYNTHESIS WITH ENOLATE EQUIVALENTS

The formation of new C–C bonds can also be achieved by using enolate equivalents. Examples of these kinds of reactions are given in the following reaction
schemes.

1. **Alkylation of enolates**: making C2–C3 bonds.
   One equivalent of a base (selected with pKa 5 units higher than acidic
   reactant) is used to ensure full alpha-hydrogen deprotonation, therefore no
   self-condensation by-products, only reaction is with electrophile.

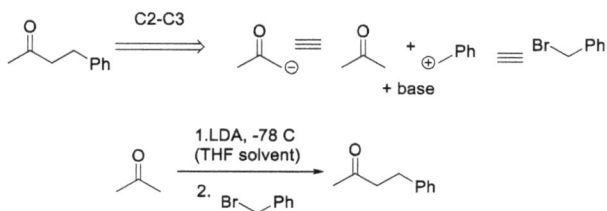

Alternatively, esters can be used as a synthetic equivalent to enolates; as the alpha H has a lower pKa, so a weaker base can be used (not NaOH to avoid ester hydrolysis), e.g. NaOMe/NaOEt, to match the ester. See also the use of enamines.

Avoid LDA with aldehydes, which are too electrophilic, leading to undesired products from self-condensation. Instead, protect as a silyl enol ether, activated with MeLi.

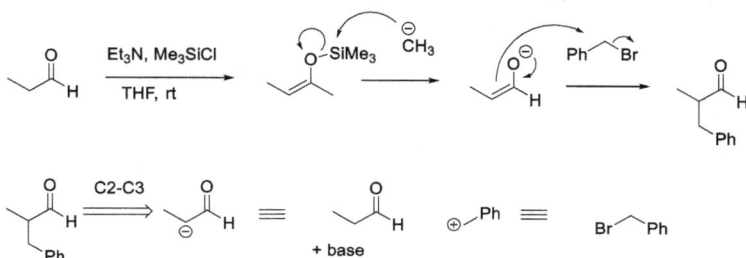

Note on regioselectivity: ketones have enolisable protons on either side of the carbonyl, so for unsymmetrical ketones, two regioisomers are possible.

i.   Kinetic control: Large LDA molecule will attack the least sterically hindered site of the ketone.

ii. Thermodynamic control: using silyl enol ether, the most stable enolate is made, i.e. most substituted alkene.

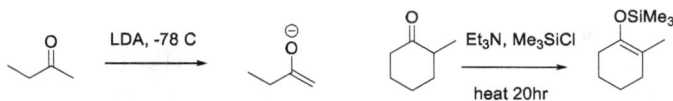

2. Synthesis of 1, 3-difunctional compounds

For 1, 3-dicarbonyl compounds, there are two options for a C2–C3 disconnection. For example, look at the intramolecular Claisen condensation (Dieckmann cyclisation).

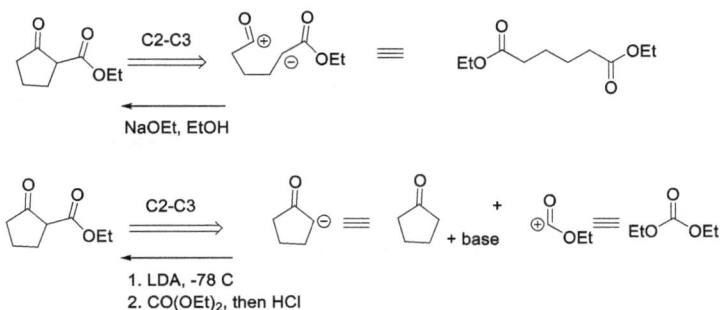

3. Synthesis of 1, 5-difunctional compounds

The target molecule has two C=O groups in 1, 5 relation to each other. Refer to Regioselective reactions of α, β-unsaturated carbonyls.

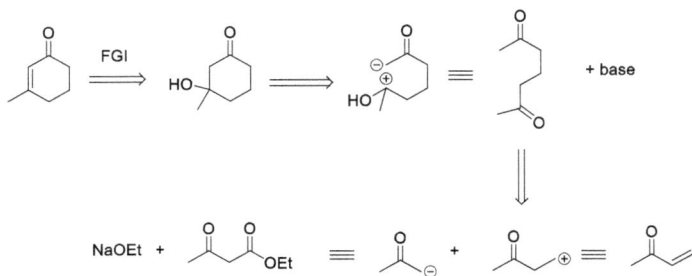

NaOEt +

Synthesis

1. NaOEt
2.

NaOEt, heat

1. NaOH
2. HCl
heat

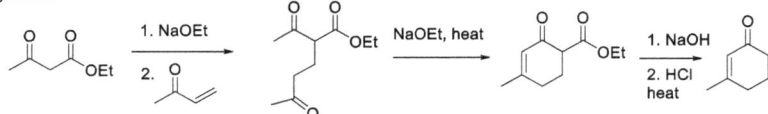

3.1.7 Umpolung (Reverse Polarity Reactions)

The synthetic equivalents for umpolung reactions allow for a reversal of the normal reaction profile: this increases the synthetic opportunities for making C–C bonds.

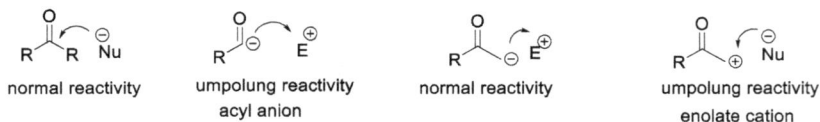

normal reactivity

umpolung reactivity
acyl anion

normal reactivity

umpolung reactivity
enolate cation

1. Synthetic equivalents of acyl anions

   1, 3-Dithiane ions are synthetic equivalents of acyl anions and are formed from the reaction of aldehydes (or ketones) with propane-1, 3-dithiol. The anion formed on deprotonation of dithiane reacts with a range of electrophiles, such as halogenoalkanes, carbonyls and epoxides, and the dithiane can be converted into an aldehyde or ketone.

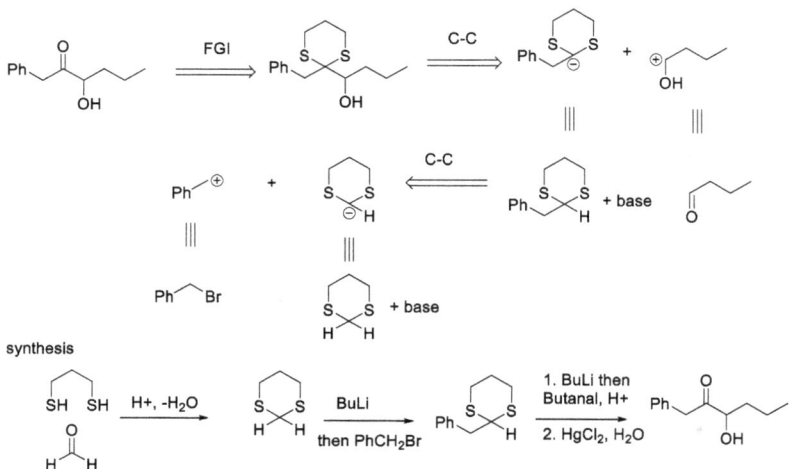

FGI

C-C

C-C

+ base

+ base

synthesis

SH   SH   H+, -H2O

BuLi

then PhCH2Br

1. BuLi then
   Butanal, H+
2. HgCl2, H2O

2. Synthetic equivalents of enolate cations

Alpha-bromo carbonyls and ketene thioacetals are synthetic equivalents of enolate cations.

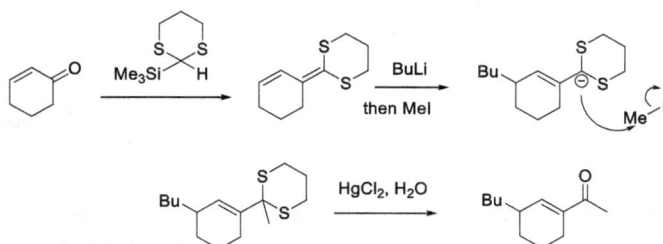

3.1.8 RING FORMATION: THE DIELS–ALDER REACTION

The Diels–Alder reaction forms two C–C bonds and a six-membered ring, with control over stereochemistry, in a single step. Hence these kinds of reactions, such as those shown in the reaction schemes below, are useful in synthesis. Dienes need to adopt the s-cis conformation. Normally the diene is electron-rich and the dienophile electron-poor to ensure good overlap of the dienophile and diene molecular orbitals. The reaction is stereospecific, a cis dienophile gives a cis configured product. The less stable endo product is preferred in irreversible reactions (endo rule); the exo product is preferred in reversible reactions.

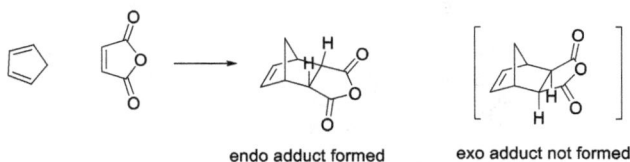

endo adduct formed   exo adduct not formed

The reaction is regiospecific, ortho- and para-substituted products are favoured (orbital interactions).

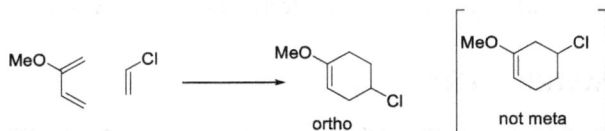

ortho   not meta

Asymmetric intermolecular Diels–Alder reactions of achiral reagents form racemic products. A chiral auxiliary can be used to make enantiomerically pure products. A Lewis acid lowers dienophile LUMO and coordinates to dicarbonyl to hinder rotation around N–CO, thereby increasing rate.

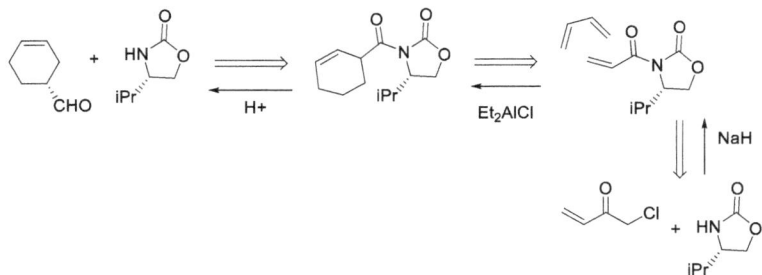

Intramolecular Diels–Alder reactions are useful for the simultaneous construction of two rings. Advantages of IMDA are close proximity of reacting groups, resulting in higher effective concentration and lower entropy of activation. This allows for milder reaction conditions and fewer degrees of freedom in the transition state, leading to higher selectivity.

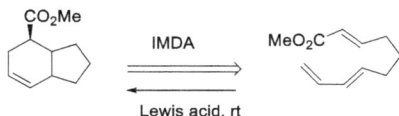

In tethered reactions, a removable tether can temporarily link a diene with a dienophile. A tether needs to be easily introduced, stable to key reaction conditions, easily removed, e.g. silicon tether. This also allows stereospecificity.

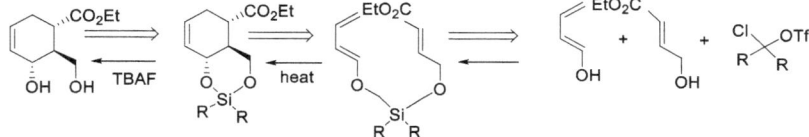

Polymer-supported reactions can be done where the diene is attached to a solid polymer support and then reacted with excess dienophile (and Lewis acid) to drive the reaction to completion. The product is then detached from the solid support, e.g. with TBAF and isolated by filtration (c.f. Merrifield peptide synthesis).

3.2   ASYMMETRIC SYNTHESIS

Regioselectivity is important in organic synthesis. There are a number of techniques available to ensure control over which isomers are formed during an organic reaction. These methods can be applied during an **asymmetric synthesis** to produce mainly one desired isomer in preference over the undesired isomer.

3.2.1 DIASTEREOSELECTIVE ADDITION OF NUCLEOPHILES TO C=O

1. Cram chelation control

The addition of a Grignard reagent to a ketone that has a stereogenic centre alpha to the carbonyl is a highly diastereoselective reaction, preferring one alcohol isomer. This is due to two reasons:

- Conformational fix: the starting material has a conformational preference, which arises from having a heteroatom on the alpha carbon that can donate a pair of electrons and the choice of nucleophile, where chelation occurs between Mg^{2+} and the oxygen lone pairs.
- Steric hindrance: one the conformation of the ketone is fixed, then steric factors control which face of the C=O is preferentially attacked.

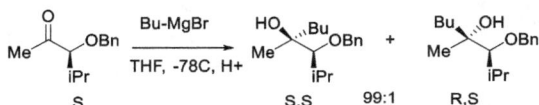

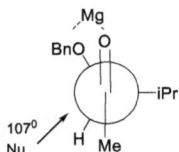

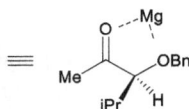

We can see from the Newman projection, which represents the diastereomeric transition state, that attack along the Burgi-Dunitz trajectory is most easily done across the H rather than iPr, due to steric hindrance, i.e. lower activation energy.

2. Felkin–Anh model

Diastereoselectivity may occur even in the absence of an electronegative heteroatom (chelating group) at the alpha stereogenic centre, albeit less yielding. Conformational fix arises from placing the largest group perpendicular to the C=O; this is the lowest energy conformation, being least sterically hindered. The nucleophile will attack along the Burgi–Dunitz trajectory from the face opposite the large phenyl group over the smallest group in the favoured transition state.

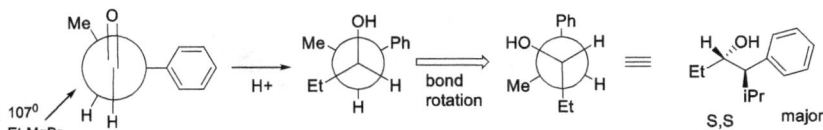

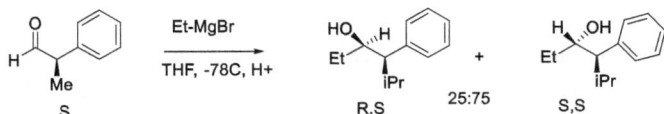

3. Modified Felkin–Anh model

For instance, where the chiral ketone does have an electronegative atom at the alpha stereogenic centre, but there is no chelating metal, placing Cl perpendicular to the C=O fixes the conformation, as electrons from the

Cl–C bond can donate into the C=O bond and the nucleophile adds opposite to the electronegative atom, over the smallest group.

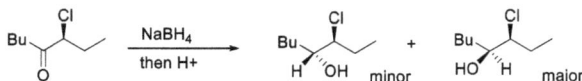

3.2.2 Asymmetric Reduction of Ketones Using the CBS Catalyst

Reduction of a ketone readily generates a chiral alcohol. Diastereoselectivity is achieved through using a chiral catalyst. See predictive mnemonic:

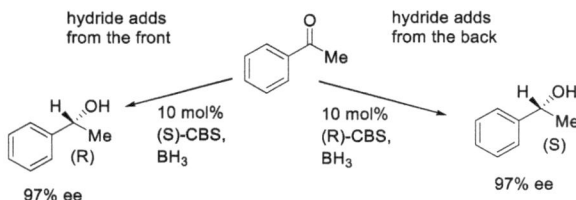

- Ketone must be in this orientation, i.e. with the V pointing up, the large group pointing down on the left and the small group pointing down on the right.

The amine of the CBS catalyst donates its electrons to BH_3 to give a very reactive boron hydride-like species that will carry out the reduction. Coordination of the ketone cis to the BH_3 brings it in close proximity, forming a diastereomeric transition state. Conformational fix comes from the chair conformation of the six-membered TS, then steric factors dictate that the largest ketone substituent (Ph) prefers to be in an equatorial position; the small group (Me) will be axial.

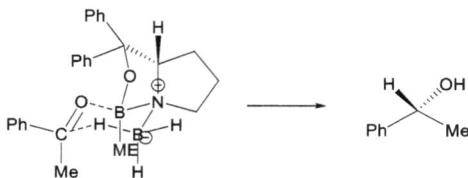

CBS reduction of ketones is useful in synthesis because one enantiomer of alcohol can be selectively produced. Activation of the alcohol into a good leaving group, e.g. tosylation, means that S_N2 can be used to introduce other groups, such as amines and sulphides (NB: there will be a change of configuration).

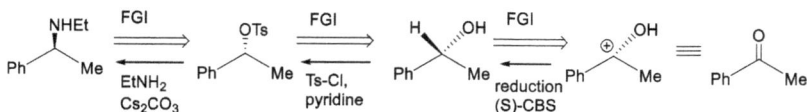

3.2.3   ASYMMETRIC α-ALKYLATION OF ENOLATES USING OXAZOLIDINONE

Diastereoselective α-alkylation of an enolate forms a C2–C3 bond to produce one stereoisomer by using a chiral auxiliary, oxazolidinone.

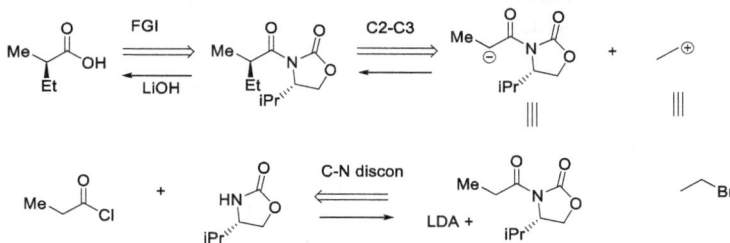

The three key features of the transition state are that the enolate geometry has the R group and oxygen cis. The conformational fix is chelation of the lithium enolate to the C=O of the oxazolidinone. Steric hindrance causes electrophile to attack opposite the sterically bulky group (iso-Pr) on the chiral auxiliary. See predictive mnemonic:

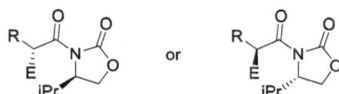

- The V of the RCH$_2$C=O must be drawn pointing down and the two C=O groups pointing up. The electrophile attacks opposite the iPr group on the chiral auxiliary.

3.2.4   ASYMMETRIC SYN-ALDOL REACTIONS USING OXAZOLIDINONES

This reaction uses a boron enolate and oxazolidinone as a chiral auxiliary, but the conformation fix is different to α-alkylation. The key part of the mechanism is that the R1 group in the alpha-position to C=O will be cis to OB(Bu)$_2$ in the boron enolate to avoid steric hindrance, hence it is always the syn aldol product. Coordination of the aldehyde to the boron enolate gives a six-membered TS, called a Zimmerman–Traxler transition state. Conformation fix comes from the dipoles of the boron enolate and oxazolidinone being arranged in opposite directions. To minimise steric hindrance, the aldehyde coordinates opposite the chiral auxiliary, with the R2 group equatorial.

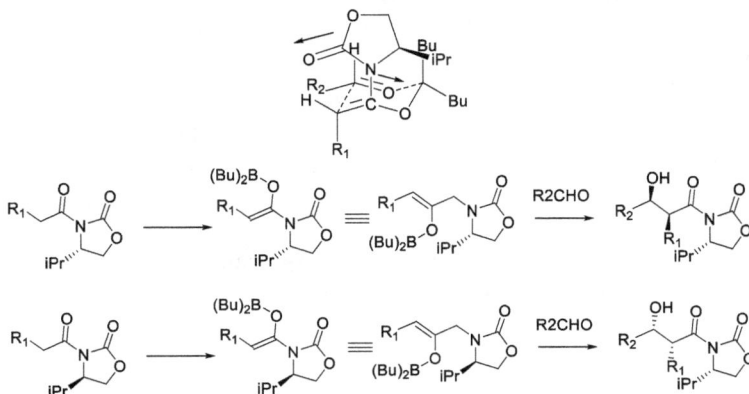

- **Predictive mnemonic**: starting material and boron enolate are drawn with the V of the R1CH$_2$C=O pointing down and the two C=O groups pointing up. The syn aldol product forms with the carbon zigzag chain drawn in this orientation, then the R1 and the iPr group will be opposite and the OH and R1 group will be syn.

For use in synthesis, look at the target molecule, identify the aldol disconnection and check that it is a syn aldol target molecule. Then decide on which enantiomer oxazolidinone is required.

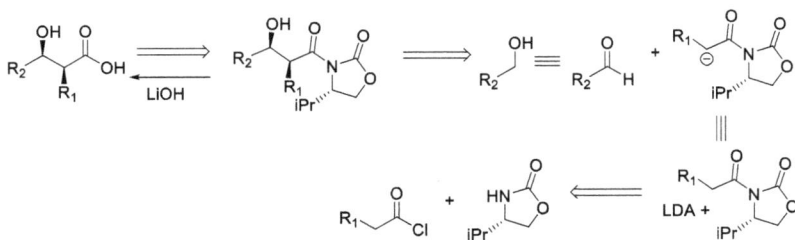

3.2.5   Catalytic Asymmetric Anti-aldol Reactions Using Proline

C–C bond construction can be achieved through the anti-selective aldol reaction between two different aldehydes mixed together with 10 mol% (S)-proline amino acid catalyst. The mechanism involves formation of an enamine with the least sterically hindered aldehyde.

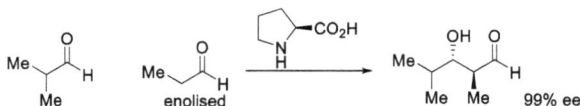

Again, the three things that control diastereoselectivity are enamine geometry, conformational fix and steric hindrance. The enamine geometry has the amine and Me group trans to each other. The conformation fix comes from two hydrogen bonds to the carboxylic acid proton on the (S)-proline catalyst: (1) intramolecular hydrogen bond to the amine, (2) intermolecular hydrogen bond to the oxygen of the aldehyde. As a result, six atoms in a ring form a chair TS. Steric hindrance places the large group (iPr) equatorially.

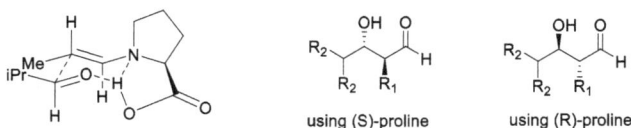

using (S)-proline   using (R)-proline

- **Predictive mnemonic**: as long as the zigzag chain of the aldol product is drawn in exactly this orientation, these enantiomers of the anti-aldol product will be created.

3.2.6 Catalytic Asymmetric Dihydroxylation of Alkenes

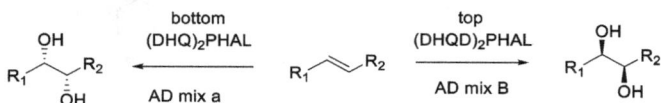

- **Predictive mnemonic**: as long as the zigzag chain of the alkene and product are drawn in exactly this orientation, the $(DHQ)_2PHAL$ attacks from the bottom and $(DHQD)_2PHAL$ attacks from the top. May need to re-orientate alkene onto this mnemonic.

3.3 STEREO-CONTROLLED SYNTHESIS USING ORGANO-MAIN GROUP REAGENTS

Organo-main group reagents can be utilised for the stereo-controlled synthesis of organic compounds, which show reaction schemes utilising a variety of reagents for the stereo-controlled synthesis of organic compounds.

3.3.1 Stereo-controlled Formation of Alkenes

1. **Formation of E-alkenes via the Julia reaction**: the use of sulphone in a two-step alkene forming reaction. To avoid the harsh conditions, the Kocienski modification uses a tetrazole functionalised sulphone.

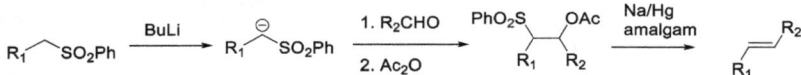

2. **Formation of E or Z alkenes via the Peterson reaction**: the formation of alkenes from β-hydroxy silanes with acid or base, where stereochemistry of the β-hydroxy silane and conditions control geometry of the alkene formed.

acid promoted reaction

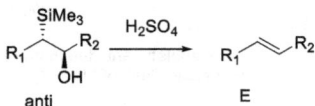

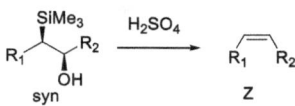

base promoted reaction

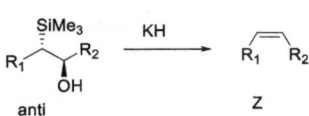

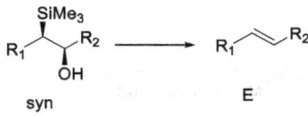

3. **Addition of ally boranes to aldehydes**: as the lone pairs on the carbonyl oxygen can donate into the empty p-orbital, a closed cyclic transition state is possible, giving 1,2-stereocontrol.

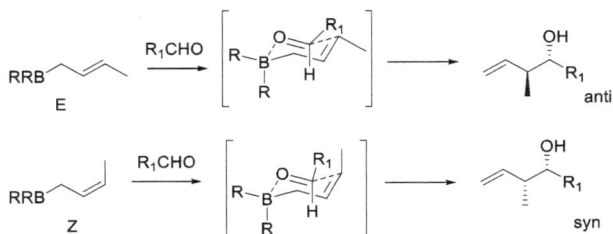

3.3.2 Control of 1,2-Stereocentres in Acyclic Systems

We can use boron enolates to generate aldol products directly, via Zimmerman–Traxler TS, and with control over the syn/anti-diastereoselectivity.

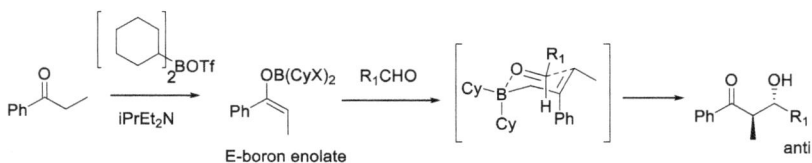

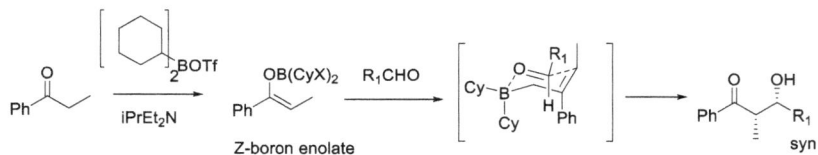

1. **Formation of 1,2,3-stereotriads**: if the aldehyde has a α-stereocentre, how will that affect addition of E and Z enolates and the product stereochemistry?

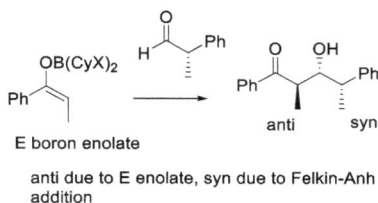

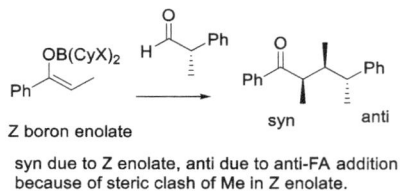

E boron enolate

anti due to E enolate, syn due to Felkin-Anh addition

Z boron enolate

syn due to Z enolate, anti due to anti-FA addition because of steric clash of Me in Z enolate.

2. **1,3-stereocontrol**

formation of anti 1,3-diols

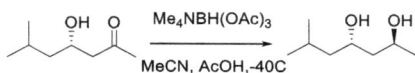

Me$_4$NBH(OAc)$_3$

MeCN, AcOH,-40C

formation of syn 1,3-diols

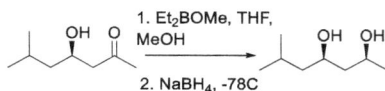

1. Et$_2$BOMe, THF, MeOH

2. NaBH$_4$, -78C

3.3.3 AXIAL VS EQUATORIAL ATTACK IN CYCLOHEXANE SYSTEMS

Axial attack is less hindered for small nucleophiles (NaBH₄, CN). Larger nucleophiles will suffer 1,3-diaxial interactions with highlighted protons, hence come in equatorially.

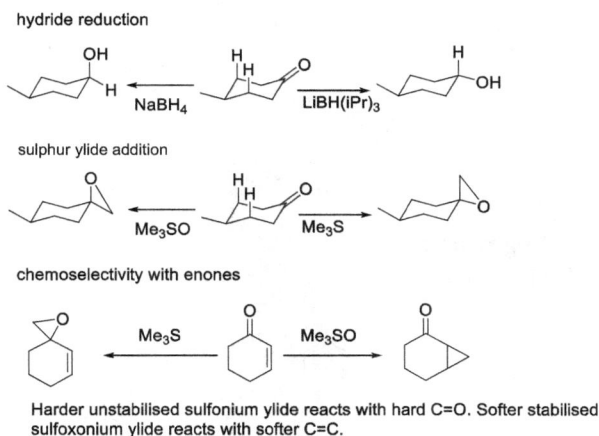

hydride reduction

sulphur ylide addition

chemoselectivity with enones

Harder unstabilised sulfonium ylide reacts with hard C=O. Softer stabilised sulfoxonium ylide reacts with softer C=C.

3.4 METAL-MEDIATED SYNTHESIS

The use of transition metal species for advanced synthesis offers a diverse range of chemical reactions that can be used for creating target compounds. Examples are given below, where reagents containing different transition metals are utilised. The structure and bonding of transition metals, with their vacant d-orbitals, mean that they are able to participate in organic reaction mechanisms in their own unique way, offering scope to organic chemists for synthesis.

3.4.1 CROSS-COUPLING CATALYSIS

1. **Catalytic process**

Nucleophilic substitution at sp² hybridised carbon is made routine by using transition metal-mediated catalysis. Classification of the different reactions is based on the main group metal used to transfer R1 in the trans-metallation event.

Catalyst activation requires Pd with vacant coordination sites, in the zero-oxidation state. Stable starting materials, such as $PdCl_2(PPh_3)_2$, need to be reduced. Reduction is assisted with an organometallic or a base, e.g. tertiary amine. See catalytic cycle:

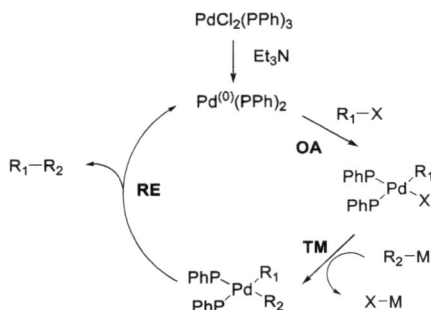

2. **Classic coupling reactions**

i. Kumada cross-coupling

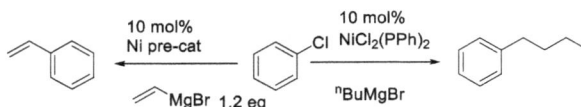

ii. Negishi cross-coupling

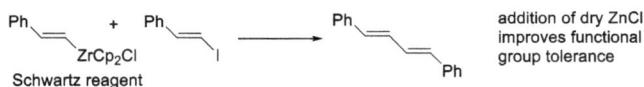

addition of dry ZnCl
improves functional
group tolerance

iii. Stille cross-coupling

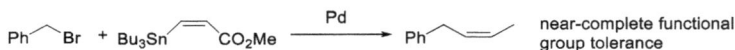

near-complete functional
group tolerance

iv. Suzuki–Miyaura cross-coupling

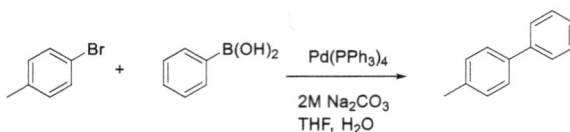

3. **C–H bond activation processes**

Organometallic reagents are often difficult to handle and must be produced in situ, e.g. by replacing R–M with R–H:

v. Sonogashira cross-coupling

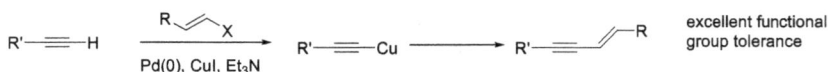

excellent functional
group tolerance

Alternatively, C–H functionalisation can be done through a β-hydride
elimination catalytic cycle. However, this has less scope in synthesis
because R–X bond must be to an sp^2 carbon.

vi. Heck reaction

Many of the reactants are pre-functionalised and need to be prepared separately. The most atom-efficient route uses reactants where neither is pre-activated.

3.4.2 ALKENE METATHESIS

1. Ring closing metathesis

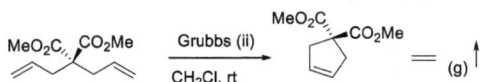

- Reactant concentration (typically 0.001–0.005 M, dilute) is critical for successful RCM macrocyclisation reactions.
- Interactions that increase the rigidity of the substrate (Thorpe–Ingold effect) and reduce entropic cost of cyclisation can be beneficial in RCM.
- Remote substituents can influence reaction efficacy and alkene stereoselectivity (e.g. OR OH) which can partially coordinate to the metal catalyst.

2. Cross metathesis

In cross metathesis, it is essential to thwart the homodimerisation of each alkene. The steric bulk in one alkene can assist in favouring the cross reaction, hence the necessity to protect the alcohol.

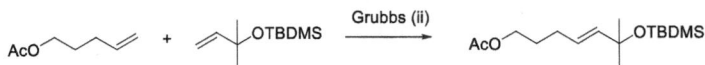

3.4.3 PAUSON–KHAND REACTIONS

catalytic P-K example

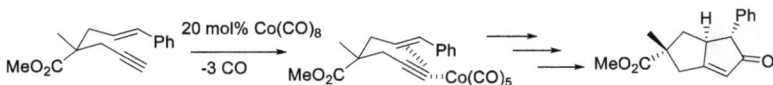

stoichiometric P-K example

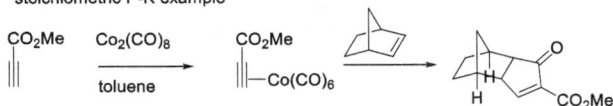

Generally, mono-substituted terminal acetylenes will give P–K products where the substituent is found in the alpha position of the enone (regiochemistry dominated by steric effects). For disubstituted acetylenes, an electron donating substituent tends to prefer the alpha position of the enone, whereas an electron-withdrawing group prefers the beta position (electronic effects). The catalytic reaction only works intramolecularly.

3.5 HETEROAROMATIC COMPOUNDS

Benzene and substituted benzenes react via the electrophilic aromatic substitution mechanism unlike alkenes, which react with electrophiles via an addition mechanism. This is on account of the aromaticity of the ring system. That is, the delocalised π-system of electron, which gives aromatic compounds their unique properties. Aromatic rings may also contain hetero atoms, such as nitrogen. These compounds will also react with electrophiles.

3.5.1 FIVE-MEMBERED AROMATIC HETEROCYCLES

The aromaticity of the rings depends on the electronegativity of the heteroatom; reluctance to donate lone pair of p electrons, but also due to atomic radius, as a larger heteroatom relieves strain in the planar ring. Hence the order of aromaticity: benzene > thiophene > pyrrole > furan.

Like benzene, aromatic heterocycles will react with electrophiles. The ability to push electrons around the ring from the heteroatom makes them very nucleophilic. Attack by the electrophiles takes place at C2/C5 in preference to C3/C4 (regioselectivity) because there are more resonance structures available to stabilise the positive charge.

Thiophene will react with a range of electrophiles, but pyrrole and furan are acid sensitive, e.g. alternative nitration.

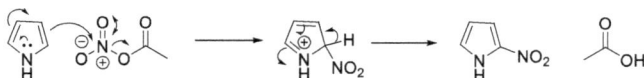

e.g. Mannich reaction.

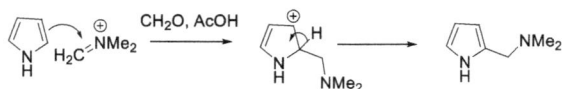

Reaction with bases can produce useful nucleophiles for synthesis.

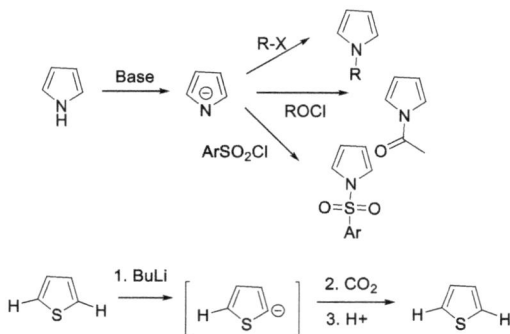

3.5.2 Six-Membered Aromatic Heterocycles

In pyridine, the nitrogen atom is sp^2 hybridised as well as the carbons. The lone pair of electrons is in an sp^2 orbital and is not part of the aromatic sextet; hence there are 6 π electrons over six centres, but the ring is electron deficient, due to the electron-withdrawing effects of N. Pyridine is therefore less reactive than pyrrole, but can undergo a wider range of reactions.

1. **Reactions at the nitrogen atom**

    The availability of the lone pair of electrons, which are not part of the aromatic system, means that pyridine can act as a base in protonation and form adducts with Lewis acids. Also, N-alkylation/acylation can occur.

2. **Electrophilic substitution**

    The mechanism isn't a simple electrophilic aromatic substitution, but instead undergoes an addition–elimination via a pyridinium salt. It requires harsh conditions and/or an activating substituent on pyridine. Attach at C3 is favoured because this avoids placing positive charge on the nitrogen through resonance.

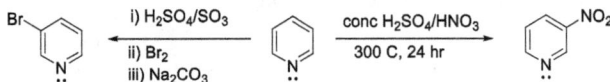

3. **Nucleophilic substitution**

    Pyridine readily undergoes nucleophilic substitution at C2 and C4, as nitrogen acquires a negative charge through resonance. This is most easily done by displacing a halogen from the ring rather than hydrogen. (Note that substituents can be added at C2/C4 by electrophilic substitution by activation as pyridine N-oxides.)

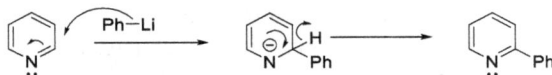

4. **Methyl pyridine**

    The electron-withdrawing effect of the nitrogen makes the methyl group acidic, so is a useful intermediate in synthesis.

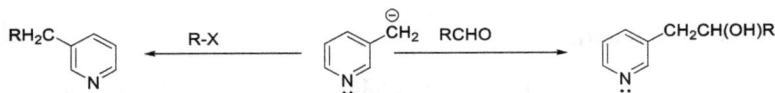

3.5.3 FUSED RING SYSTEMS

Multiple aromatic ring systems may be fused together, and the delocalised character of the π-orbital is spread across the entire ring system. The reactions demonstrate how fused ring systems can be used in synthesis. Many natural products and drug compounds contain aromatic fused ring systems.

1. **Quinolones and isoquinolines**

   A benzene ring can be fused to a pyridine ring in two ways:

   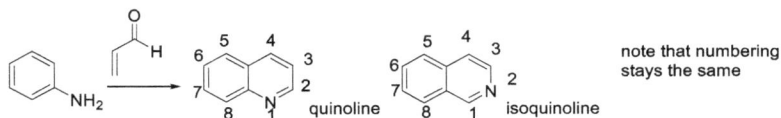

   i.   Electrophilic substitution:

       In acidic conditions, nitrogen is protonated. Substitution takes place on the benzene ring which is more reactive than pyridine, attacking at C5 to retain aromaticity.

   ii. Nucleophilic substitution:

       These reactions are similar to pyridine, preferably displacing a halogen.

2. **Indoles**

   Reactions occur at C3 to retain aromaticity or the pyrrole nitrogen can act as an acid to become a reactive anion.

   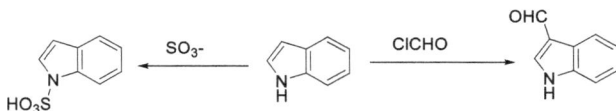

3.6 SYNTHESIS OF NITROGEN-CONTAINING COMPOUNDS

Nitrogen occurs commonly in organic molecules both in nature and in the laboratory. It constitutes several important functional groups, such as amines and amides, and is present in heteroaromatic ring systems. Nitrogen-containing functional groups will undergo a range of reactions. The lone pair on nitrogen means that amines can act as nucleophiles, while amides will react like other carbonyls via electrophilic substitution. These compounds are important in nature, being present in many biomolecules, such as protein and nucleic acids, so are highly relevant to medicinal chemistry. Hence, effective ways to synthesise nitrogen-containing compounds are essential to medicinal chemistry.

3.6.1 AMIDE FORMATION

Amides are commonly found in pharmaceuticals, so we need reliable ways to synthesise them.

i.

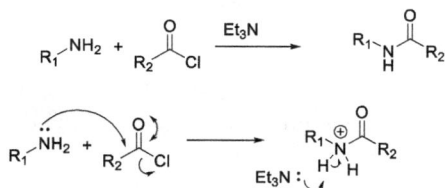

ii.

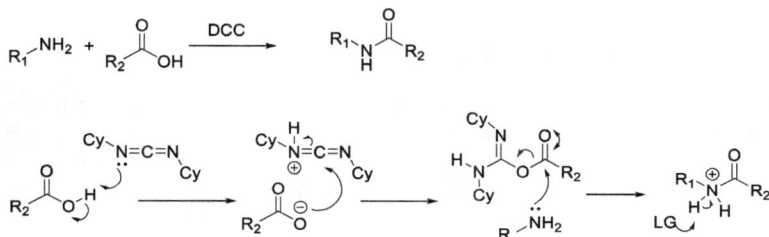

RSA of amides is a very useful disconnection.

e.g.1

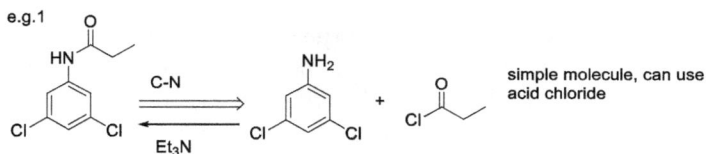

simple molecule, can use
acid chloride

e.g.2

acid used as carbonyl; hard
to make acid chloride from
complicated molecule.

The other N lone pair is
delocalised into the benzene
ring, therefore less nucleophilic.

3.6.2 REDUCTIVE AMINATION

Reductive amination is a widely used method to synthesise amines.

less reactive reducing agent than
NaBH$_4$ due to EWG to avoid reduction
of ketone/ aldehyde or acitic acid.

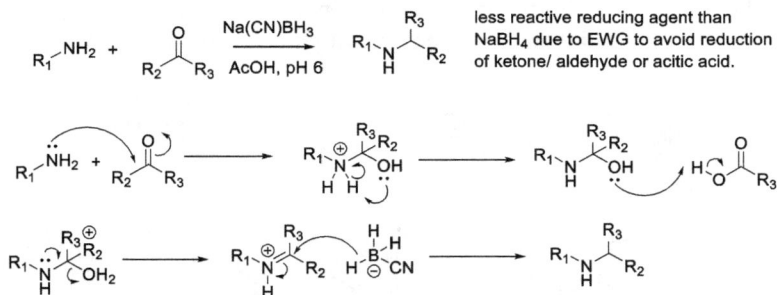

Acetic acid is a weak acid that gives the solution pH 6, needed for the reaction to work:

- pH < 4: amine is protonated mostly and will not act as a nucleophile.
- pH > 7: no acid is present to protonate the alcohol to make a good LG.

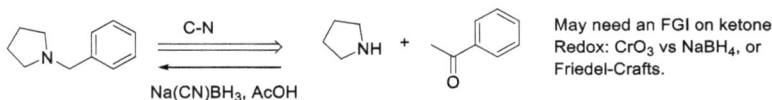

3.6.3 AMINATION OF ALKYL HALIDES

An amine can be introduced to a molecule through S_N2 reaction with an alkyl halide, as shown in the mechanism below. Note inversion of stereochemistry.

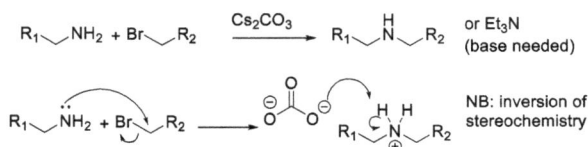

The alkyl halide may be Cl, Br or I, primary or secondary, but not tertiary. The amine can be primary or secondary, but more substituted amines may lead to over-reaction, being more nucleophilic (+I effect of alkyl group), to form a tertiary amine product, therefore use sterically hindered groups if possible. Note that S_N2 is impossible for aryl halides – nucleophile can't approach from the back.

Alternatively, alkyl sulphonates (e.g. tosylates) can be used in an S_N2 alkylation. Choose to use tosylates where the leaving group is attached to a stereogenic centre – there are many ways of synthesising stereogenic alcohols, but not alkyl halides.

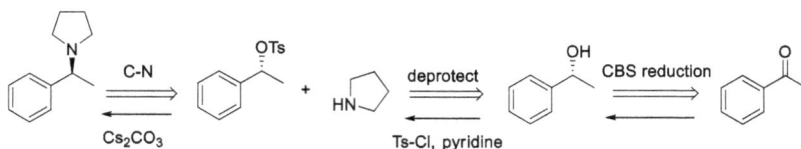

3.6.4 AMINATION OF AROMATIC HALIDES

Nucleophilic aromatic substitution (addition–elimination mechanism) is possible for benzyl halides, provided there is an electron-withdrawing group on the ring.

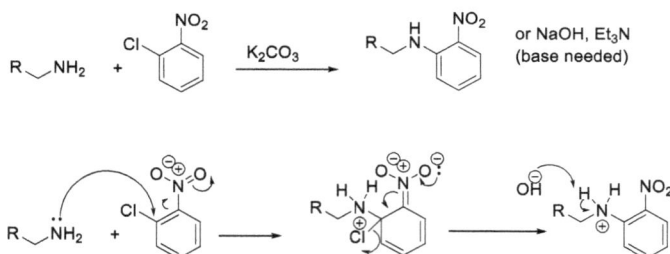

The reaction is best with more electronegative halide leaving group, F, Cl. Primary or secondary amines can be used. Other electron-withdrawing groups can be used, such as SO_2R, CN, COR, but must be ortho or para to the leaving group – conjugation.

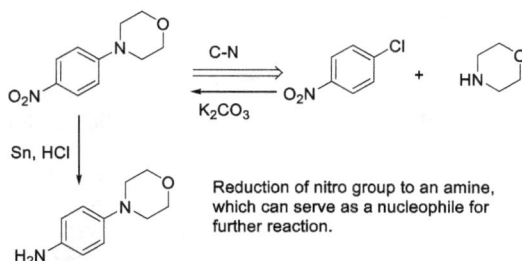

Reduction of nitro group to an amine, which can serve as a nucleophile for further reaction.

The addition–elimination mechanism is also possible for 2- and 4-halopyridines because the nitrogen in the ring serves as the electron-withdrawing group; can also use ROH and other nucleophiles to attack pyridine/benzyl halides.

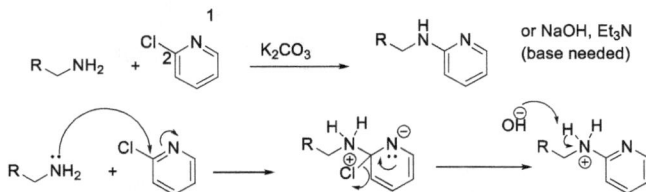

3.6.5 AMINATION OF AROMATIC HALIDES WITH TRANSITION METALS

1. **Pd-mediated coupling is a frequently used disconnection in medicinal target molecules**

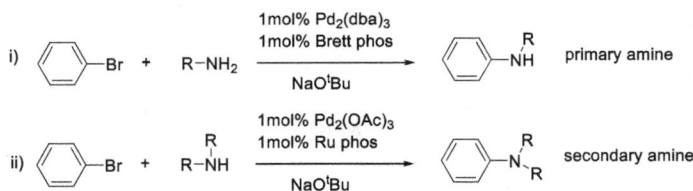

This works well for all aromatic bromides/iodides and types of amine. Unlike S_NAr mechanism, which requires an EWG on the aromatic halide, Buchwald–Hartwig amination works on both electron-rich and electron-poor aromatic halides, so is much more versatile.

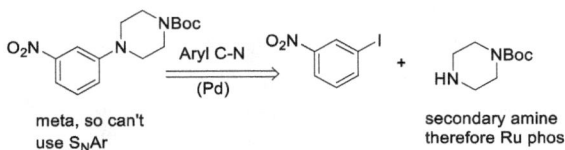

meta, so can't use S_NAr

secondary amine therefore Ru phos

2. Cu-mediated coupling: amidation of aromatic halides

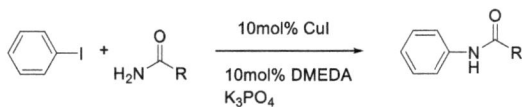

This works well for all aromatic iodides (electron-rich–electron-poor) and primary amides. Best method to synthesise cyclic amides (lactams) – no other good method of making them.

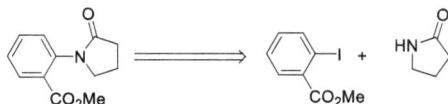

3. Suzuki-aryl couplings using Pd

The aryl–aryl motif is very common in medicinal chemistry, used to include pyridines also.

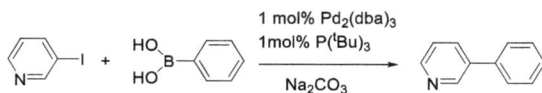

3.6.6   α-Functionalisation of Cyclic Amines

Racemic deprotonation and trapping is a useful way to introduce groups in the position next to the nitrogen atom – deprotonate Boc-protected cyclic amines with a strong organolithium base, then use an electrophile to trap. ^{S}BuLi is a very strong base; attacks protons that are not very acidic (high pKa) TMEDA coordinates to Li, making the base even more reactive.

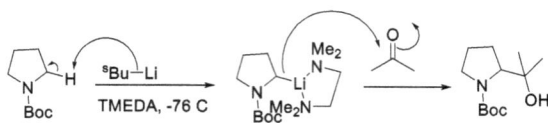

α-Functionalisation via Negishi reaction is used to introduce an aromatic ring.

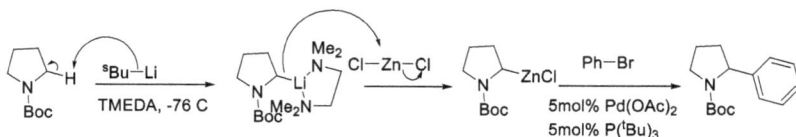

Use of chiral diamines in place of TMEDA can give stereoselective α-functionalisation of N-Boc cyclic amines. The chiral diamine coordinates with the ^{S}BuLi to make a chiral base which can preferentially remove one of two enantiotopic protons to give a covalent sp^3 hybridised organolithium intermediate which is trapped with retention of stereochemistry.

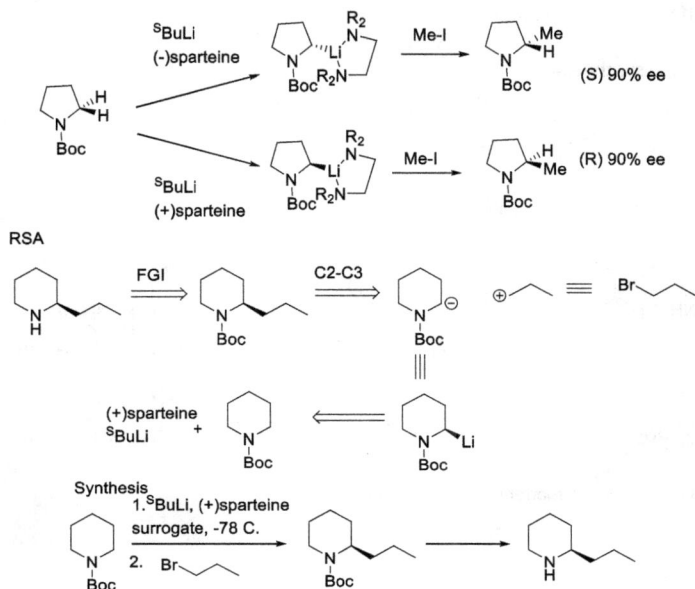

3.6.7 SYNTHESIS OF HETEROAROMATICS

Heteroaromatic molecules can be synthesised using amines, as shown in the reaction schemes below. These nitrogen-containing fused ring systems are very common in natural products and pharmaceutical compounds.

i. Fischer indole synthesis

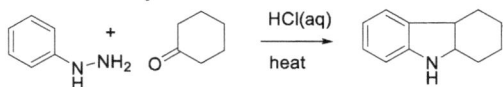

i) Fischer indole synthesis

stage 1: formation of iminium ion and enamine

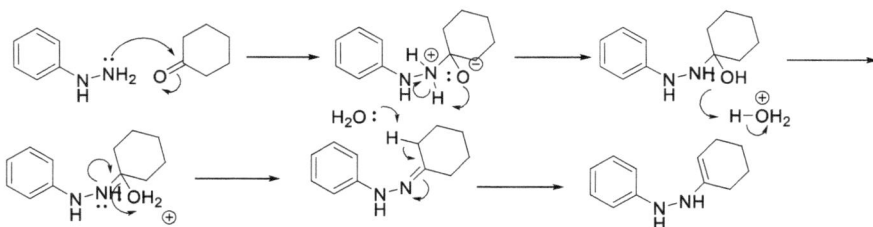

stage 2: [3,3]-sigmatropic rearrangement to form C-C bond.

stage 3: rearomatisation and los of NH₃

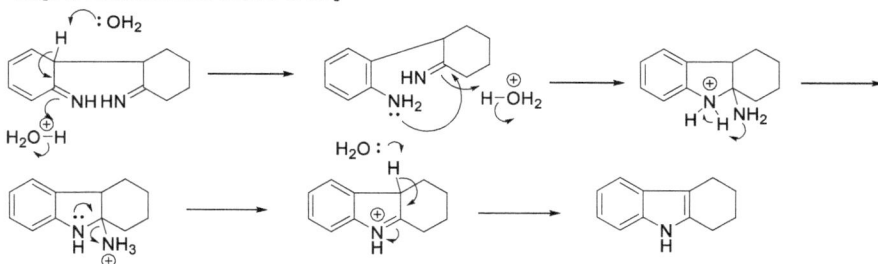

There are two key aspects of regioselectivity to consider:

1. Aromatic hydrazines must only have one position to react (can be achieved by symmetry), otherwise a mixture of regioisomers will be produced.

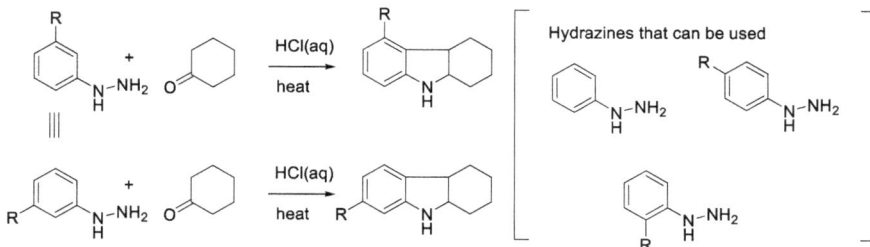

Hydrazines that can be used

2. Ketones must also be symmetrical or only have one position where enamine can form. This can be achieved by using aldehydes as well.

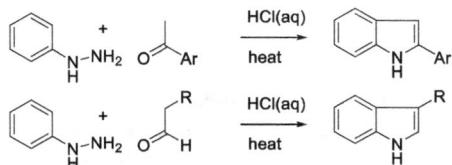

RSA – disconnect indole, ensuring regiochemistry is selective.

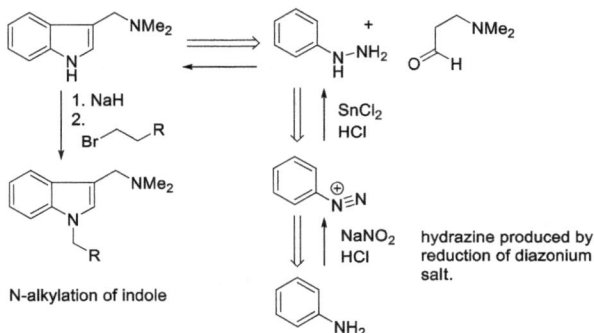

1. NaH
2. Br~~~R

N-alkylation of indole

SnCl₂
HCl

NaNO₂
HCl

hydrazine produced by reduction of diazonium salt.

ii. Benzimidazole synthesis

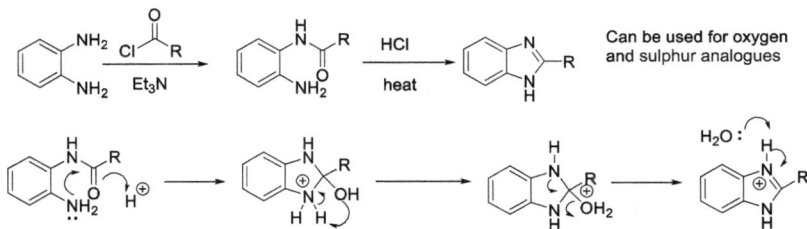

Can be used for oxygen and sulphur analogues

iii. Tetrazole synthesis

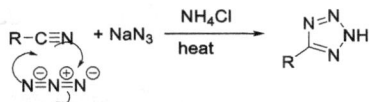

$R-C\equiv N$ + NaN₃ $\xrightarrow[\text{heat}]{\text{NH}_4\text{Cl}}$

3.7   RADICALS IN SYNTHESIS

Reactions involving radicals, such as those with their mechanisms shown below, can be very useful in synthesis. Radical reactions involve the homolytic cleavage of a chemical bond, where one electron from the pair that formed the covalent bond is shared between each of the cleaved species. The movement of these electrons is shown using 'fishhook' or single-sided curly arrows. There are different types of radical reaction, and they provide alternative means of synthesising new molecules.

3.7.1 TYPES OF RADICAL REACTIONS

1. **Radical abstraction (S_H2 substitution, homolytic, bimolecular)**
   More stable product, faster reaction (relative rate: R_2 = tertiary > secondary > primary). Driving force of the reaction is formation of strong H–OR, H–Cl bonds.

Intermolecular example - hydrogen abstraction

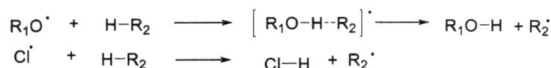

$$R_1O^{\cdot} \ + \ H{-}R_2 \ \longrightarrow \ \left[R_1O{-}H{\cdot\cdot}R_2 \right]^{\cdot} \longrightarrow \ R_1O{-}H \ + R_2^{\cdot}$$

$$Cl^{\cdot} \ + \ H{-}R_2 \ \longrightarrow \ Cl{-}H \ + R_2^{\cdot}$$

Intramolecular example - 1,5 atom transfer

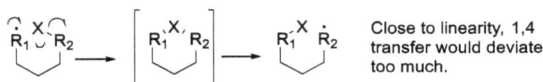

Close to linearity, 1,4 transfer would deviate too much.

2. **Radical addition**
   Radicals add regioselectively to unsymmetrical C=C bonds (e.g. the peroxide-catalysed anti-Markovnikov addition of HBr to alkenes, preferential attack to least sterically hindered carbon forms secondary radical as major product – intermolecular example).

Intramolecular example – useful for making five- and six-membered rings.

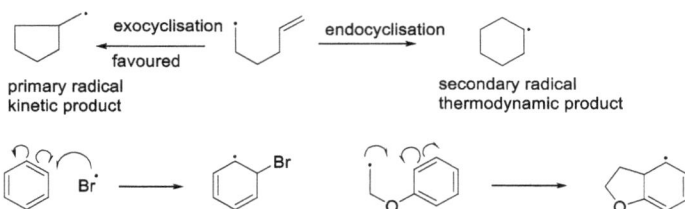

exocyclisation favoured

endocyclisation

primary radical
kinetic product

secondary radical
thermodynamic product

3. **Radical fragmentation (reverse of radical addition)**
   Driving force of the reaction is either breaking weak C–X bond (X = R_3Sn, RS, Br) in place of a stronger bond or entropically favourable products.

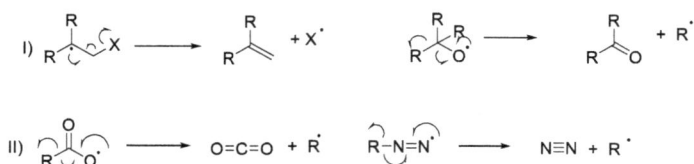

I)

II)

4. Termination reactions

  i   Recombination or coupling – reverse of bond homolysis

$$\dot{R} + \dot{R} \longrightarrow \quad R–R \quad \text{dimerisation}$$

  ii   Disproportionation – transfer of a hydrogen atom

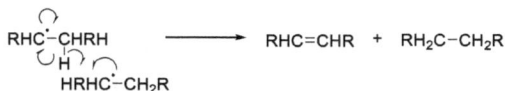

$$RH\dot{C}–CHRH \longrightarrow RHC=CHR + RH_2C–CH_2R$$
$$HRH\dot{C}–CH_2R$$

  iii   Electron transfer

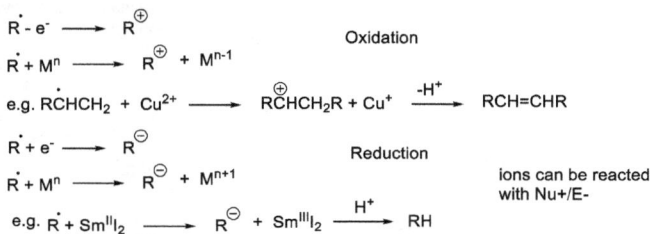

$$\dot{R} - e^- \longrightarrow R^{\oplus} \qquad \text{Oxidation}$$

$$\dot{R} + M^n \longrightarrow R^{\oplus} + M^{n-1}$$

$$\text{e.g. } R\dot{C}HCH_2 + Cu^{2+} \longrightarrow R\overset{\oplus}{C}HCH_2R + Cu^+ \xrightarrow{-H^+} RCH=CHR$$

$$\dot{R} + e^- \longrightarrow R^{\ominus} \qquad \text{Reduction}$$

$$\dot{R} + M^n \longrightarrow R^{\ominus} + M^{n+1} \qquad \text{ions can be reacted with Nu+/E-}$$

$$\text{e.g. } \dot{R} + Sm^{II}I_2 \longrightarrow R^{\ominus} + Sm^{III}I_2 \xrightarrow{H^+} RH$$

- – Radicals adjacent to an EDG have a high-energy SOMO and are called nucleophilic radicals because they donate an electron and form a cation.
- – Radicals adjacent to an EWG have a low-energy SOMO and are called electrophilic radicals; they accept an electron to form an anion.

3.7.2 INTRAMOLECULAR CYCLISATION REACTIONS

The use of tributyltin hydride in radical synthesis is advantageous because mild conditions are used, solvation is less important and the reaction is selective, so can synthesise sterically hindered polar compounds, without the need for protecting groups for strong O–H and N–H bonds.

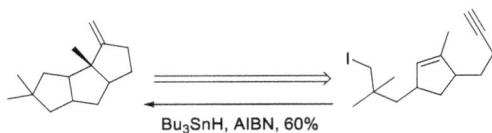

Bu₃SnH, AIBN, 60%

Initiation

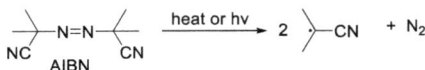

heat or hv

2 ⟩–CN + N₂

Propagation

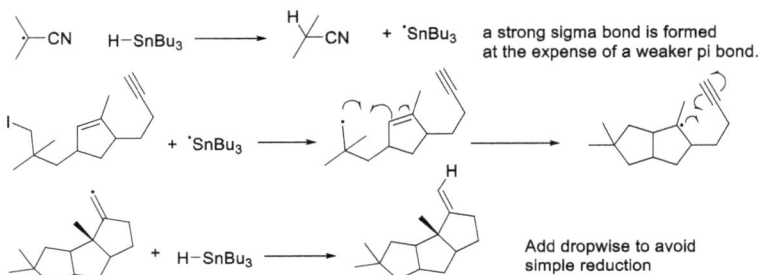

a strong sigma bond is formed
at the expense of a weaker pi bond.

Add dropwise to avoid
simple reduction

Baldwin's rules are a series of guidelines for predicting the relative favourability of cyclisation reactions to form three- to seven-membered rings, based on the proper geometries for orbital overlap and reaction. (Exo = bond broken is outside the ring being formed, Endo = bond broken is inside the ring being formed.) Geometry at the reaction centre is either tet (sp³), trig (sp²) or dig (sp). Examples:

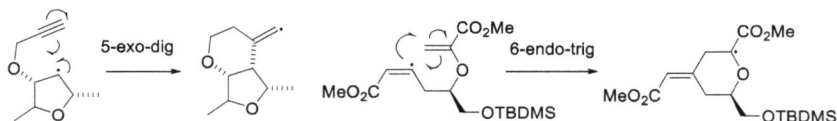

3.7.3 INTERMOLECULAR REACTIONS

For example, the synthesis of prostaglandin F2a. A C–C σ bond is formed at the expense of a weaker π bond. Bu₃SnH is added over 1.5 hours and 10 eq of ketone is used to minimise simple reduction. Addition to the C=C is regioselective. The first formed carbon radical on homolysis of C–I is nucleophilic and so it adds quickly to the electron-poor C=C bond.

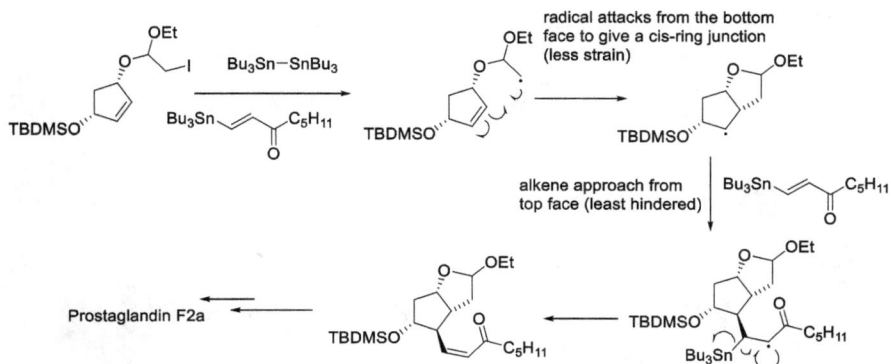

Functional group interconversions: Bu$_3$SnH reduces other functional groups besides organohalides, e.g. deoxygenation of alcohols via xanthates, RO(C=S)SR.

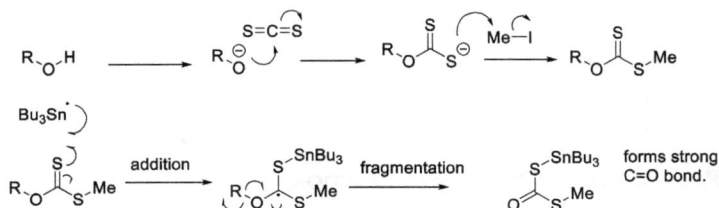

3.7.4 1,5 Hydrogen Atom Transfer

This reaction involves the movement of a hydrogen atom from its site of generation to another atom located five atoms away, following an energetically downhill pathway. The transfer allows access to radicals that might otherwise be difficult to prepare.

i. Using tributyltin hydride

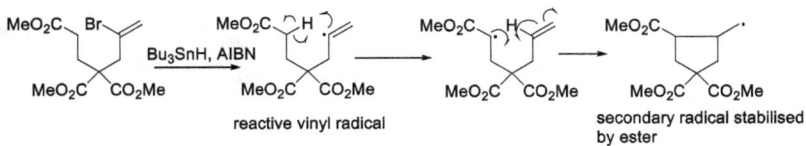

ii. Using photolysis

Carbonyls

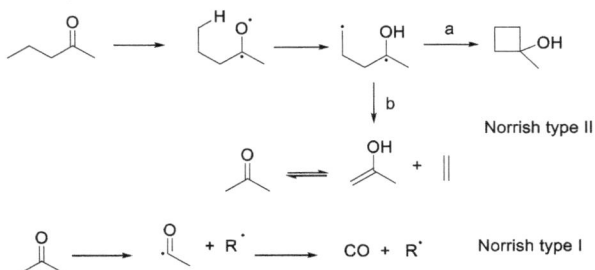

Norrish type II

Norrish type I

Nitrites: the Barton reaction

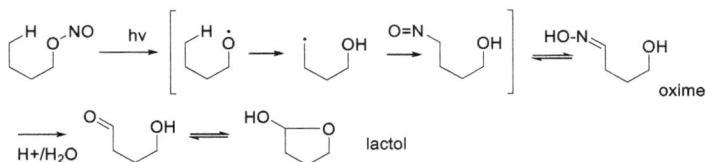

oxime

lactol

3.7.5   RADICAL POLAR CROSS-OVER REACTIONS

Radical oxidation

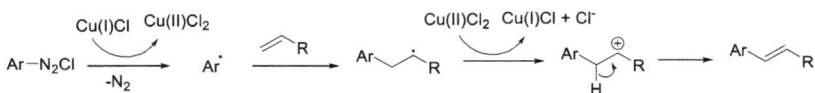

Radical reduction

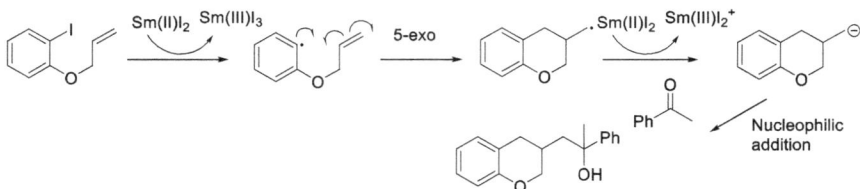

Nucleophilic
addition

TUTORIAL QUESTIONS

1. Provide a retrosynthetic analysis for the synthesis of a generic R-COCl acid chloride from an alcohol.

2. Amides can be synthesised from either acid chlorides or carboxylic acids. Suggest a retrosynthetic analysis, including reagents, for each of these two processes.

3. Show bond disconnections and suggest reagents for four aromatic reactions commonly used during synthesis.

4. The Wittig reaction can be used to make alkenes from alcohols. Comment on the different reagents required during the synthesis of a trans-alkene compared to a cis-alkene and illustrate your answer with a retrosynthetic analysis for each.

5. The aldol condensation can be used for making new C2–C3 bonds and therefore is very useful in synthesis. Give a retrosynthetic analysis for the production of a generic enone, $CH_3COCHCHR^2$, from readily available starting materials, including the appropriate bond disconnections in your answer.

4 Biochemistry of the Cell and Drug Development

ABSTRACT

Biomolecules are of crucial importance for the structure and function of the cell. Proteins are made from chains of amino acids, which fold up to form three-dimensional tertiary structures. The structure will determine the role of the protein in the cell, for example, the protein might be required for a structural purpose, or it might be needed for the catalysis of a biochemical process as an enzyme. Enzymes are critical for the catalysis of biochemical processes, which occur at the relatively mild condition of body temperature. The instructions for synthesising proteins are contained in genes on DNA. The DNA in the cell nucleus is transcribed onto RNA before being translated into a protein in the cytoplasm at ribosomes, organelles specifically designed for this process. The cell membrane and those of the organelles within the cell are made from phospholipids. These membranes allow for the controlled passage of molecular feedstocks for synthesis in and out of the cell. They also permit cellular signalling processes.

Drugs must interact with the body to produce a biological response. On a microscopic level, the tissues which comprise our bodies are made up of individual cells. It is with the cells that drugs perform their function. The cell is a very complex structure and offers numerous targets on which drugs can work. The cell likewise can be considered simply as a reaction vessel; living creatures are composed of chemicals and are obedient to the same laws of chemistry as any laboratory experiment. Knowledge of the principles of chemistry means that medicinal chemists can predict how molecules will interact with a biological target and design drugs that will generate the required response to alleviate the symptoms of a disease. The cell contains thousands of essential molecules, where chemical energy drives their biosynthetic reactions to produce the cell's fundamental components. Small molecules, which are predominantly obtained from the diet, are used to synthesise the giant macromolecules of the cell. Principles of thermodynamics give order to these polymers in this chemical system, resulting in them adopting predetermined conformations, dependent on the sequence of their monomer sub-units. These macromolecules assemble to form the vital structures of the cell, such as receptors, transport proteins, enzymes as well as non-protein structures such as plasma membranes and nucleic acids. It is within this ensemble of molecules that a molecular target for a particular disease must be identified and a drug designed to interact with this target in such a way as to serve as a therapy.

The elements that are prevalent in organic chemistry are indeed the fundamental building blocks of the cell. Nature carries out its organic reactions within the aqueous environment inside the cell, with water serving as a solvent. Polar molecules are dissolved; those with functional groups containing electronegative atoms that pull

DOI: 10.1201/9781003516835-4

the electron density of the molecule towards themselves are held within the aqueous medium, while non-polar hydrophobic compounds remain separate from the internal cellular solution, known as the cytoplasm. This property of different solubility of biological molecules is crucial to the cell, for it governs how molecules interact and react together and how the cell membrane is formed to give the cell its structure and stability. Many of the functional groups commonly used in organic chemistry are also frequently seen in nature: methyl (CH_3), hydroxyl (OH), carboxyl (COOH) and amino (NH_2) recur repeatedly in biology. The small organic molecules found in the cell generally contain up to 30 atoms and have many uses in the cell, such as intermediates for deriving energy from food and units to build polymers which comprise the majority of the cell's composition.

4.1   BIOLOGICAL MOLECULES

The synthesis and breakdown of biopolymers follow a discrete sequence of chemical changes and follow definite rules. As a result, many of the biological compounds that make up the cell are chemically related and can be broadly classified into four major families of small organic molecules: the simple sugars, the fatty acids, the amino acids and the nucleotides. Sugars are the food molecules of the cell, broken down to create chemical energy. Fatty acids play an important role as the main component of the cell membrane, and amino acids and nucleotides are the sub-units of two crucial groups of biopolymers: proteins and nucleic acids, respectively.

Biological molecules may consist of many thousands or even millions of these sub-units to produce structures where each atom is precisely linked into a specific spatial arrangement, imperative to determine the macromolecule's properties, which in turn governs their specific function in the cell. The specific sequence by which these sub-units, be it amino acids or nucleotides, are organised carries specific information and generates a biological message that can be 'read' through interactions with other molecules. This is how biological macromolecules perform their function: by interacting with the appropriate molecular counterparts, governed by specific intermolecular forces that exist between them, macromolecules can perform roles in the cell such as catalysing chemical transformations, assembly into multi-molecular structures, generate motion and most fundamental: to store hereditary information.

The cell is an aqueous environment, where water comprises 70% of the total mass. The remainder is mainly due to macromolecules. These are assembled from their monomer sub-units through particular biochemical mechanisms which specifically control the sequence of the monomers added to the end of the polymer chain as well as determining the appropriate time to terminate the sequence. The macromolecular chain is linked by covalent bonds, which are strong enough to preserve the sequence for a long period of time. This precise sequence dictates the information contained within the macromolecule, but utilisation of the information is controlled through non-covalent interactions. These are much weaker bonds which exist between the polar functional groups between different molecules or different parts of the same molecule. These electrostatic interactions govern the 3D shape of a macromolecule and therefore how it will interact with other molecules.

These interactions singularly are too weak to withstand thermal motions, hence macromolecular structures are held in place by multiple non-covalent interactions. For this to occur, spatial arrangements of the atoms must be precisely matched for a strong interaction. An example is the case of hydrogen bonding, which is a particular kind of non-covalent interaction between a lone pair of electrons on a sufficiently electronegative atom, namely nitrogen, oxygen or fluorine, and the electron-deficient hydrogen atom bonded to the aforementioned elements. The donated lone pair of electrons must be in the correct orientation relative to the accepting hydrogen atom of the other molecule/other part of the same molecule to satisfy the directionality of the intermolecular force. In biochemistry, usually multiples of these interactions are required between molecules, which leads to the need for interacting molecules to be of specific shapes, as in the case of enzyme–substrate complementarity. These exacting requirements account for the specificity of biological recognition.

The number of possible 3D shapes of a macromolecule is restricted by the nature of atoms behaving as hard spheres, with a definite radius, meaning no two atoms can overlap, which restricts the number of possible bond angles in a polypeptide chain. This in turn, and in addition to steric factors where atoms hinder one another in a congested region of a molecule, puts constraints on the number of 3D arrangements, or conformations, that the molecule can exhibit. In the case of proteins, steric hindrance from amino acid side chains and the way in which the electrons of the covalent bonds resonate between the amino acids produce a rigid primary structure, where the amino acid side chains are held pointing outwards in particular positions depending on the sequence of amino acids. This feature means that the polymer chain will fold into a predetermined secondary and, ultimately, tertiary structure. Here nature works to ensure a specific molecule is made, needed for a specific function. If the primary structure had the full flexibility expected for single covalent bonds, the synthesis of proteins would be much more erroneous.

4.1.1 PROTEINS

Proteins are built from a repertoire of 20 amino acids; all are necessary for protein synthesis, but only 10 of the 20 can be synthesised by humans; the other 10 must be obtained from dietary sources. Protein chains, often thousands of amino acids long, fold into unique three-dimensional structures due to electrostatic interactions within the chain or between other chains to produce the specific structure that governs the function of the protein. Proteins are incredibly important from a biological perspective: they have many functions in the body, attributable to their versatility. Their name, from the Greek *proteios* meaning of first rank, named in 1838 by Jöns Jacob Berzelius, indicates how crucial these molecules are to the cell because without them, cells would lose their structure, biochemical reactions would not proceed and cellular communication could not occur. Proteins are classified as either fibrous or globular, depending on this 3D structure. Proteins that are involved in the control of biochemical processes tend to be globular, while proteins involved in a mechanical occupation are usually fibrous.

Amino acids are di-functional, containing a basic amine group and acidic carboxyl group. Each of the 20 amino acids is an alpha-amino acid, that is, the amine

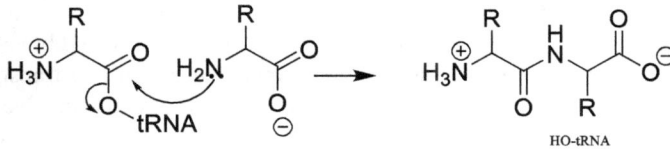

FIGURE 4.1   Overall process for the formation of a peptide bond in the cell via a protease-catalysed reaction involving tRNA.

substituent is attached to the alpha carbon next to the carbonyl group. Also note that 19 of those amino acids are primary amino acids and only differ in the side chain substituent. Having both acidic and basic groups, amino acids are able to undergo internal acid–base reactions to produce dipolar ions, called zwitterions at a particular pH, known as the isoelectric point, and are amphoteric, acting as both an acid and a base, which is crucial for their involvement in biological catalysis. A polypeptide is synthesised from amino acids by the condensation reaction between the carboxylic acid group of one amino acid with the amino group of the next to form an amide linkage, or peptide bond, via an enzyme-controlled pathway (Figure 4.1). The precise sequence of amino acids, referred to as the **primary structure**, will determine the final form of the protein due to the specific hydrogen bond formation that occurs between the nitrogen and oxygen groups of different parts of the chain.

Proteins have three-dimensional folded structures, the shape of which is very specific, depending on the almost infinite variety of amino acid sequences, and is predetermining for their given function. Proteins can fold into a regularly repeating **secondary structure**: either an α-helix or a β-pleated sheet. These structures are governed by the primary structure as hydrogen bonds are formed between different amino acids in the chain, or covalent bonds in disulphide bridges between cysteines. In an **α-helix**, the primary chain twists into a coil about a central axis, while an undulating sheet-like structure is obtained in the formation of a **β-sheet**: which of these different structures is formed depends on the precise hydrogen bonding that arises from the given sequence of amino acids. Further folding may then occur to produce a tertiary structure. For proteins composed of several constituent polypeptide chains, the final folded structure is referred to as the quaternary structure. These unique shapes have discrete implications regarding the function of the protein.

The proteins that bind together to produce a **quaternary structure**, which is the finished functional protein, are organised in a particular arrangement to produce a structure with the correct properties to perform a specific role in the cell. For example, haemoglobin carried in red blood cells consists of four sub-unit peptides, each with a cofactor, called the haem group, with an iron atom centre to which an oxygen molecule can bind. Each haemoglobin molecule can therefore carry four molecules of oxygen and the precise arrangement of the four polypeptide sub-units is crucial for the conformational changes that the protein undergoes to facilitate oxygen binding.

Half the dry mass of a cell is made up of proteins, which are responsible for the structural integrity of the cell as well as catalysis and molecular recognition. Proteins are of crucial importance in the body, having many key functions: enzymatic catalysis, transport and storage of small molecules, coordinated motion (make up muscles), mechanical support (collagen, a fibrous protein, in skin and bone), immune protection (antibodies are highly specific proteins), generation and transmission of nerve impulses and control of growth and differentiation of cells (involved in DNA replication). The information for making these proteins is stored as genes on DNA.

The instructions for producing proteins are encoded on genes, sections of DNA, which are transcribed on to RNA before being translated into a protein chain at ribosomes, cell organelles designed for this operation. This is a very intricate enzyme-controlled biological pathway, with a high degree of complexity. Proteins produced from this process then adopt their specific 3D shape and move to their intended location to perform a given function in the body. Nucleotide triplets code for particular amino acids; the order in which these triplets are assembled in the gene determines the sequence of amino acids in the protein and therefore ultimately governs the protein's shape and function. The different side chains of each of the 20 amino acids can form a variety of hydrogen bonds: hence the observed variety of possible structures. Proteins are very versatile molecules that can have many functions, governed by their structural organisation. For globular proteins such as those involved in catalysis and molecular recognition, the final folded 3D shape of a polypeptide, the tertiary structure, may constitute one protein domain, and several globular units comprise the functional quaternary structure, as with haemoglobin. In order for polypeptides to assemble into a functioning quaternary structure, precise interactions between these molecules must be in place.

Intermolecular interactions are of fundamental importance in biochemistry. The forces that exist between molecules are responsible for the assembly of protein subunits to form the quaternary structure. Intermolecular forces are crucial for biological catalysis with enzymes, molecular recognition via substrate–receptor binding and DNA processes as well as many other biochemical processes. In order to bind, the molecules must come into contact to enable formation of the non-covalent interactions. Molecules move randomly due to thermal motion as they collide, and bounce off each other. These collisions must have sufficiently low energy to permit a binding interaction to occur between molecules, i.e. collide with less energy than the enthalpy of bond formation for the intermolecular force. As such, the rate of complex formation between two molecules is said to be diffusion-limited. Diffusion is slower for large molecules with a larger collision cross-section than it is for small molecules. Furthermore, the rate of complex formation is impeded by the need for the binding surfaces to be orientated so that the interacting surfaces can fit together. This is associated with a reorganisation energy; hence most collisions do not result in bond formation. These dynamics need to be considered in the development of pharmaceuticals to assess the potency of a drug. This is referred to as pharmacokinetics. Also, understanding the thermodynamics of biochemical processes, such as protein–DNA interactions, substrate–receptor binding and enzyme–substrate complex formation, is a necessity for designing an effective drug to work on these kinds of targets. The strength of the binding is measured with an equilibrium constant $K = [HS]/[H][S]$, which gives the ratio of the concentration of the complex in solution over the concentrations of the separate host and substrate. In other words, the larger the equilibrium constant, the more host–substrate complex there is in solution, the stronger the binding. The strength of binding has important implications for drug design, for example, a drug working on an enzyme host will have a potency that depends on the strength of the binding, which can be modified through understanding of the thermo-dynamics of molecular interactions.

4.1.2 ENZYMES AND CATALYSIS

The rate at which chemical reactions proceed in the cell is exceedingly fast. Many cellular reactions are catalysed by enzymes, needed as many reactions would oth-erwise proceed at imperceptible rates at body temperature. They can catalyse bio-logical reactions on the order 10^3–10^6 reactions per second. The interactions between enzymes and substrates are governed by thermodynamics and are diffusion-limited, but the rate of reaction is enhanced by providing an alternative route for the reaction to proceed with a lower activation energy. This is achieved by binding of the enzyme, which can distort the substrate molecule into a more reactive conformation, and the kinetic aspect is improved by holding one of the reacting molecules, the substrate, in place, in the correct orientation for reaction with a second molecule.

For proteins that function as a catalyst, namely enzymes, the conformation of the protein governs its chemistry. The precise folded structure gives a unique sur-face to the protein, where neighbouring residues interact in such a way as to alter the chemical reactivity of selected amino acid side chains. Particular residues may be orientated in such a way as to optimise binding interactions with a substrate, forming complementary hydrogen bonds. Furthermore, neighbouring parts of the polypeptide chain may interact in a way to exclude water molecules from the active site, which would compete with the substrate for hydrogen bonds. While it may be difficult to conceive that a structure can restrict access to a small molecule like water, it is often energetically unfavourable for water molecules to separate from the hydrogen bond network of the aqueous medium to habituate a crevice on the protein surface. The clustering of amino acid side chains due to the particular folded structure can activate normally unreactive side chains. Negatively charged side chains can be forced together against their mutual repulsion by the tight folded structure and create an attractive binding site for positively charged substrates. When the necessary reactivity cannot be achieved by simply having the correct organisation of side chains, proteins can utilise other non-polypeptide molecules called cofactors. These bind to the protein surface and serve to exploit the reactiv-ity of these molecules that they acquire when bound to the protein; an example is the iron in haem and cytochrome C in a protein-bound metal ion complex with complicated organic chelates.

The precise shape of proteins means that enzymes catalyse specific chemi-cal reactions. This is referred to as enzyme–substrate specificity, as only certain molecules are able to bind to a given active site. This selectivity is crucial for reactivity, as enzymes speed reactions by selectively stabilising transition states. A reacting molecule adopts a high energy conformation (the transition state) and precise molecular interactions are required from the enzyme to stabilise this struc-ture so that the reaction can proceed correctly. The catalytic ability of enzymes is far superior to any synthetic catalyst, and this efficiency is due to several factors.

As previously mentioned, the enzyme increases the local concentration of the substrate at the catalytic site by holding it in the active site in a position with the correct orientation for the reaction. Most important are the binding energy contributions. Upon substrate binding, the enzyme modifies the geometry of a substrate molecule to produce an unstable structure with a different electron distribution and a much higher free energy value. This structure, the transition state, is therefore higher in energy than the substrate, being in a reactive conformation. The activation energy, the minimum amount of energy needed for the reaction to proceed, is equal to the difference in energy between the original substrate and the TS, so energy input is needed for the reaction to proceed. The reason these reactions can occur much more frequently than they otherwise would in the absence of a catalyst is because the binding energy of the enzyme lowers the energy of the transition state, bringing it closer in energy to the favoured conformation of the original substrate, hence lowering the activation energy, therefore less energy input is needed. Without enzyme intervention, much more energy would be needed to distort the substrate to the reactive conformation, therefore the reaction would be energetically less favourable and proceed at a slower rate.

The way in which enzymes promote a reaction, by breaking and making covalent bonds, can proceed via different mechanisms. In the absence of enzymes in a laboratory scenario, peptides can be hydrolysed using either an acid or a base catalyst. The precise positioning of amino acid side chains in the active site means that enzymes can have the unique ability to utilise amphoteric residues for acid and base catalysis simultaneously because the acidic and basic residues are prevented from combining with each other, as they would do in solution, due to being bound to the rigid protein framework. As a result, a specific substrate, with the precise requirements to fit the active site, will be positioned in such a way that the acidic and basic residues are in the correct position to disperse the electron density of the substrate so that electron density is withdrawn from the substrate by the acidic residue which increases susceptibility to the appropriate nucleophile, e.g. water in the case of hydrolysis. Meanwhile water is made into a stronger nucleophile by the action of the basic residue pushing electron density onto the reactive centre, in this case oxygen, resulting in the rate of reaction being increased.

The reaction profile in Figure 4.2 illustrates how the energy difference between the substrate and the transition state at the top of the curve, which is the activation energy, is lowered in the formation of an enzyme–substrate complex. Less energy is required for the reaction to proceed, so more encounters between the enzyme and the substrate will lead to a successful reaction, hence rate is increased. The mechanism for hydrolysis of a peptide bond explains how the catalytic process operates. The negative charge of the carboxylate pushes electron density onto the electronegative oxygen atom of the water molecule, making it a strong nucleophile. Meanwhile the electronegative nitrogen atom of a second amino acid residue pulls electron density from the C=O bond, creating a partial positive charge on the carbon, making it very susceptible to nucleophilic attack from the oxygen lone pair of electrons, hence the reaction proceeds quickly.

The example given above shows how proteins can be catalytically hydrolysed, for example, during digestion of food. Enzymes may also increase rates of reactions in

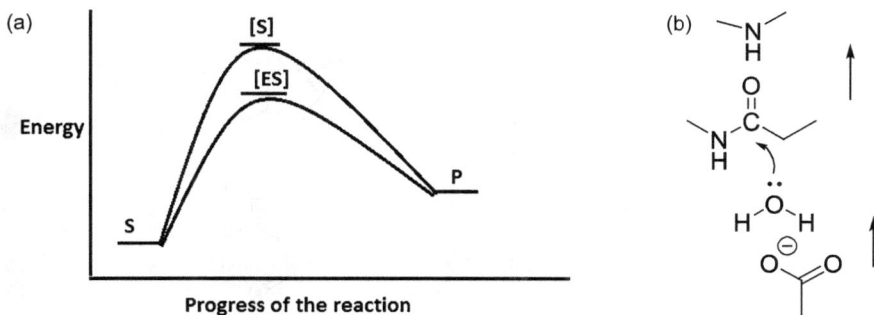

FIGURE 4.2 A schematic reaction profile is shown in diagram (a) for hydrolysis of a peptide. Diagram (b) shows the mechanism for hydrolysis of a peptide bond.

other ways. For example, they may form a covalent bond with the substrate, which therefore becomes attached to a residue or coenzyme in the active site. This can have the effect of weakening the covalent structure of the substrate, which will then react more readily with a second molecule, which causes cleavage of the bond to the enzyme, which is now free to catalyse another reaction. This is how serine proteases work, for example.

4.1.3 NUCLEIC ACIDS

The diversity of proteins required for a cell to function is transcribed from the cell's instruction manual: DNA. The DNA (deoxyribonucleic acid) resides within the cell nucleus, which is effectively the control centre for the cell. The nucleus is a membrane-bound structure containing pores to permit the selective passage of certain molecules required for cellular activity at this organelle. Nucleotides and polynucleotides, such as RNA, continually diffuse in and out of the nucleus. The code for building the proteins of the cell is carried from the control centre on ribonucleic acid (RNA) to the cytoplasm where the information contained on these molecules is interpreted and translated into an amino acid sequence to assemble a protein to be used in the cell. Protein synthesis, along with most other cellular activities, is controlled from the genes contained within DNA.

The elucidation of the structure of DNA resulted in the Nobel Prize being awarded for this discovery to James Watson and Francis Crick, thanks to the work performed by Rosalind Franklin, who used X-ray diffraction to obtain images of DNA. These images allowed the structure of DNA to be determined, as they contained important information about the structure of DNA. They showed that the structure of DNA comprises a double-stranded helix made from the hydrogen bonds between nitrogenous bases that connect the two strands of the sugar–phosphate backbone together. The structure of DNA is intrinsically linked to the function of the molecule in the cell, controlling the flow of genetic information during DNA replication (cell cycle/fertilisation) and gene expression during protein synthesis.

DNA is a polymer made up of monomer units called nucleotides. **Nucleotides** are simple molecules that are essential for life, either as a single unit or polymeric, with

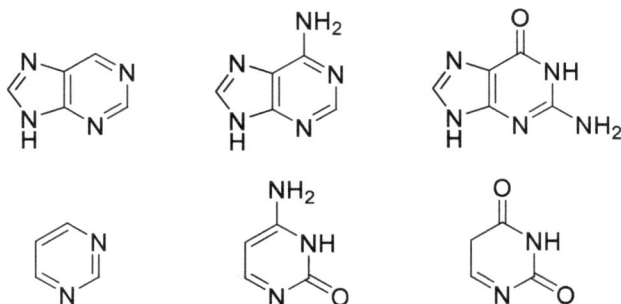

FIGURE 4.3   The molecular structures of purines and pyrimidines.

numerous roles inside the cell. They participate in nearly all biochemical processes, are required for the storage of molecular information on DNA, are vital in protein synthesis via mRNA and are involved in intracellular signalling (e.g. cAMP) and form co-enzymes (NADH) and energy-rich intermediates (ATP). The fundamental chemical building block of DNA and RNA is the nucleotide, which comprises three components: a nitrogenous base, a pentose sugar (furanose; deoxyribose/ribose) and a phosphate group to bridge between adjacent sugars within nucleic acid chains.

There are two basic types of nitrogenous bases, which are **purines** and **pyrimidines**. Both are derived from carbon and nitrogen ring structures. Purines are based on two carbon–nitrogen rings: one hexagonal and one pentagonal. Pyrimidines are based on a single hexagonal carbon–nitrogen ring. They are shown in Figure 4.3. Both of these heteroaromatic rings are flat and non-polar (hydrophobic), which is important for their role in forming the DNA helix.

The bases form hydrogen bonded pairs with other bases; however, they require defined binding partners. Adenine binds to thymine or uracil (depending on DNA or RNA) and guanine binds to cytosine. This is because purines will only bind to pyrimidines. This is due to the size of the molecular structures fitting inside the helix and the precise binding requirements of the hydrogen bonds that exist between the bases. Guanine can only bind cytosine (and vice versa) because the interaction requires three hydrogen bonds. Adenine can only bind thymine (or uracil) as the interaction requires two hydrogen bonds. Other combinations of these molecules will not bind correctly as they would repulse. This is referred to as base pair complementarity, as shown in Figure 4.4, and is a consequence of the precise positioning needed for hydrogen bond formation.

The sugar molecule in the nucleotide structure is a monosaccharide, shown in Figure 4.4. The monosaccharide can either be ribose or 2-deoxyribose. The C_2 carbon is key in identifying which: if C_2 is –H, this is 2-deoxyribose; if C_2 is –OH, this is ribose. The carbon at point 2 on the monosaccharide shows whether or not the molecule is 2-deoxyribose or ribose. 2-Deoxyribose is seen in DNA and has a –H whereas ribose has –OH and is seen in RNA. Bases always have a glycosidic linkage to the anomeric carbon of the sugar. When a sugar and a nitrogenous base bind, the structure is known as a nucleoside. Nucleotides include the phosphate group which attaches two adjacent sugar molecules by phosphodiester bonds between a 3′

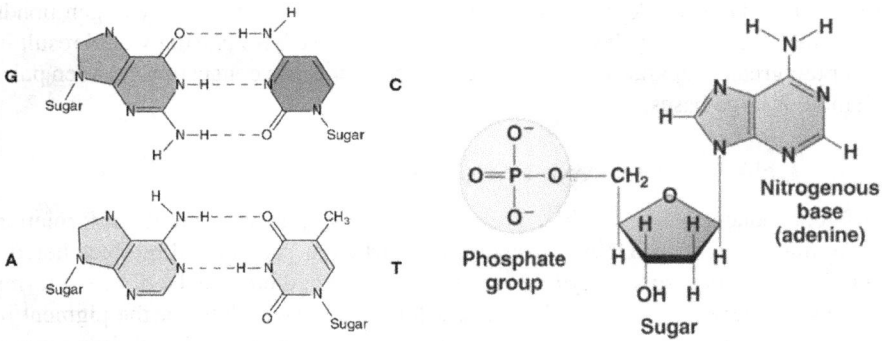

FIGURE 4.4 The hydrogen bonding between nucleotides and structure of a nucleotide.

carbon and a 5′ carbon in DNA, forming the sugar–phosphate backbone. Phosphate groups of one molecule attach to the sugar of another nucleotide. C_3 –OH forms a phosphodiester bond with C_5 –PO_4. Nucleotides are always added to the 3′ end of a polynucleotide chain. This means the strand will 'grow' in a 5′ to 3′ direction by the formation of phosphodiester bonds. This gives directionality to the chain. Phosphate groups make the nucleotide negatively charged and water soluble (via electrostatic interactions with weakly charged water molecules) and this has ramifications for the stability of DNA. The non-covalent binding interactions of the outer backbone with the aqueous medium of the cytoplasm combined with the hydrophobic interaction of the bases in the centre of the helix make the structure of DNA energetically stable.

DNA strands are antiparallel (different directions of growth). Any nucleic acid chain has a 5′ phosphate end – the start of the chain. The other end is 3′ and terminates with a –OH group. Conventionally, DNA sequence is written in the 5′ to 3′ direction – so the example is ATTGC and its complementary strand is GCAAT.

To summarise, the DNA molecule is a long unbranched polymer composed of nucleotide sub-units, which contains the bases: adenine (A), cytosine (C), guanine (G) and thymine (T). The nucleotides are linked together covalently with phosphodiester bonds between the 5′ carbon of one deoxyribose group and the 3′ carbon of the next. These linkages form the sugar–phosphate backbone of the DNA molecule to which the four bases are attached. Two strands of DNA combine via intermolecular bonding between pairs of bases, one from each strand to form the complete molecule. These base pairs are instrumental in determining the characteristic structure of DNA. Two DNA chains combine through hydrogen bonds between the bases like runs of a ladder and cause the structure to twist into the α-helix shape. This structure was discovered by Watson and Crick in the early 1950s from X-ray diffraction studies. Their model of the DNA structure revealed that the DNA bases were inside the double helix and that the bases of one strand were extremely close to those of the other, which results in the need for specific base pairing between a large purine base (A or G, which contains a large double ring structure) and a smaller pyrimidine base (T or C, each of which has a single ring) on the other chain. These requirements for spatial arrangements of atoms to satisfy hydrogen bonding result in complementary base pairing between A and T and G and C, where two and three hydrogen bonds are formed, respectively, between the bases. Alternative base pairing would result in either too greater separation for hydrogen bonding or steric congestion between pairs of small or large bases.

4.1.4 DNA REPLICATION AND PROTEIN SYNTHESIS

Different sequences of these base pairs comprise a gene. Genes are the information containing elements of DNA that code for a particular protein and produce hereditary characteristics, such as eye colour. In a population, there may be different forms of the same gene. In the example of eye colour, the gene coding for the pigment in the iris varies between people who have different coloured eyes. The different forms of a gene are called alleles. Remarkable diversity of proteins is achieved from these four bases. The unique sequence of the bases on a gene governs the specific sequence of amino acids during protein synthesis. The DNA sequence exhibits a triplet code, where a particular combination of three base pairs codes for a particular amino acid when translated from the RNA sequence at the ribosome, which was transcribed from DNA. A particular sequence of triplets codes for a specific protein.

During cell division, identical copies of these genes must be transmitted to each daughter cell. The information contained within the genes is written as the unique linear sequence of nucleotides that make up that section of the DNA. A consequence of complementary base pairing means that each strand contains identical information and can serve as a template for the replication of that information. This is the principle that underpins DNA replication and protein synthesis. Errors may occur during DNA replication and this causes mutations.

During protein synthesis, mRNA (messenger RNA) is produced by the process of transcription and is a template for protein synthesis. Cytoplasmic mRNA is subjected to translation, where a ribosome reads codons on the mRNA and creates a polypeptide chain. This process is assisted by tRNA (transfer RNA) which carries a specific amino acid (codon) and is the link between the mRNA sequence and the coded protein sequence. The tRNA forms a very unique secondary 'cloverleaf' structure, which is crucial for its function in the cell. The 3′ end is where amino acids attach. An anticodon loop is the binding site for mRNA.

All this occurs at ribosomes in the cytoplasm. The rRNA (ribosomal RNA) is the primary component of ribosomes, essential for protein translation. It carries out protein synthesis in ribosomes by forcing tRNA and mRNA to process and translate the latter into proteins. Ribosomes are 60% rRNA and 40% ribosomal proteins. They have a complex structure, specific to their role in protein synthesis.

Another kind of RNA, **miRNA** (microRNA) are regulators of cell development, growth, survival and disease via RNA silencing of mRNA. mRNA is silenced via cleavage, destabilisation and less efficient ribosomal translation. One miRNA can control and impede translation of up to 50 different mRNAs. Some miRNA are involved in disease processes. miRNA is involved in carcinogenesis as a causative agent. Half of all chronic lymphocytic leukaemia of B-lymphocytes lacked the genes to encode 2 miRNAs. miRNAs are crucial tools in cancer prognosis and diagnosis and are also becoming targets for the development of therapeutic tools.

4.1.5 LIPIDS AND THE PHOSPHOLIPID BILAYER

All the cellular processes that proceed in a controlled environment within the cytoplasm are in isolation from the disruptive influences of the external environment. This is attributable to the protective qualities of the cells' outer membrane. The plasma membrane is crucial to the cell because it maintains the internal environment within the cytoplasm and separates the operating systems of the cell from the extracellular environment, which is essential for the sensitive biochemical processes that are the habitual conduct of the cell. The plasma membrane enables selective passage of substances into and out of the cell to sustain the delicate chemical system within. Structures inside the cell, called organelles, also comprise a plasma membrane. Among these membrane-bound organelles are the Golgi apparatus, endoplasmic reticulum, mitochondria as well as the cell nucleus, all of which play a crucial role in sustaining the dynamic operating systems inside the cell. The contents within these organelles, related to their specific function, are characteristically different to the rest of the cytoplasm, maintained by the plasma membrane that keeps these systems separated.

Organelles perform different roles in the cell. The mitochondria are responsible for energy generation. Through the process of respiration, they manufacture ATP, a source of chemical energy. Modification of selected molecules is undertaken at the Golgi apparatus and/or the endoplasmic reticulum. Molecules manufactured by the cell, or assimilated from outside, may need to be changed to make them suitable for a given biochemical process. The Golgi apparatus and endoplasmic reticulum are the organelles with the capacity to perform this role. The nucleus, as previously mentioned, is where DNA is contained and is the control centre for the cell. Having a porous membrane enables passage of selected molecules to facilitate communication between the nucleus and the rest of the cell to ensure that instructions are received for the essential biochemical process of the cell to continue functioning without error.

The activities of these organelles and differentiated cells are governed by the specialised membranes. They have unique transport proteins for selective transport of molecules or ions, related to their function. Neurons, for example, rely on particular transport proteins used to conduct nerve impulses by generating a potential difference (voltage) across the cell membrane by transporting sodium and potassium ions, which gives a difference in charge between the inside of the cell and the tissue fluid outside. Mitochondria undertake aerobic respiration through which chemical energy in the form of adenosine triphosphate (ATP) is produced through biochemical processes involving the transport of hydrogen ions across specialised membranes. Release of enzymes, other protein moieties, hormones etc. tends to be assisted by transport proteins. The release of insulin from cells of the pancreas to control blood sugar is a notable example because errors that can arise in receptor signalling or transport of this small protein-based hormone can result in diabetes. The cell membranes contain specific receptors needed for a specific response related to cells' function in the tissue, such as receptors for insulin to control of blood sugar level. Receptor proteins transfer information across the cell membrane rather than ions/molecules. The cell is then able to instigate a response to this information, such as release of a hormone or neurotransmitter, initiate cell division or perform whatever response is required. Confusion in cell signalling can therefore obviously lead to problems.

Although the biological membranes have different functions, the general structure of the plasma membranes is similar. It is composed of a phospholipid bilayer in which the characteristic proteins are embedded. This consists of a very thin film which encloses the cell or organelle and is described as a 'fluid mosaic' as the lipid and protein molecules are free to move around in the plane of the membrane because the structure is held together by a dynamic array of non-covalent interactions that continually break and re-form, permitting the motion of molecules across the membranes' horizon.

The phospholipid bilayer is established due to the oil-like nature of the lipid. When oil is dropped into water, it will form a layer at the surface (provided it is less dense than water) separated by a phase boundary, which is the interface between the hydrophobic non-polar hydrocarbon chains of the oil molecules and the polar water molecules that are repelled by the oil molecules. As droplets of oil within the volume of the water ascend, they too are repelled by the water molecules and form a globule, or a hollow vesicle, due to the hydrophobic interaction. This interaction between lipid and water to form a vesicle is how a cell membrane is derived. The difference is that the cell membrane is composed of a bilayer. This originates from the nature of the phospholipids that comprise the cell membrane. They consist of a polar head, the phosphate, which interacts with the water molecules at the interface, and a hydrophobic hydrocarbon tail, which is repelled by the water, and so is buried internally within the membrane. Here the hydrophobic tails interact with mutual attraction to hold the structure together. This results in a double layer about 5 nm thick that is relatively impermeable to water-soluble molecules, so allows for selective transport of substances in and out of the cell to maintain a constant internal environment, i.e. an isolated chemical system.

The composition of the phospholipid bilayer affects the fluidity of the structure. Eukaryotic cell membranes contain especially large amounts of cholesterol, which bind to the polar head group of the phospholipid, thus reducing the mobility. Regions of the cell membrane containing large amounts of cholesterol are therefore less deformable. The higher cholesterol composition also has the effect of increasing the permeability barrier. The phospholipid bilayer is less permeable in this instance; predominantly water- and lipid-soluble small molecules are the only substances that have ease of passage. This is important for the consideration of drug design if the target is inside the cell.

Most of the specific functions carried out by the plasma membrane are conducted by membrane proteins. As such, the type and quantities of these proteins in plasma membranes vary a great deal, depending on the function of that membrane. In myelin membranes, which function as the electrical insulators in nerve cells, less than 25% of the membrane mass is protein. By contrast, the inner membranes of mitochondria, which are involved in energy transduction, are about 75% protein, by mass. Typically, cell membranes will have approximately 50% of their mass being proteins. Protein molecules are much larger than the lipid molecules that comprise the bilayer, so there are many more lipid molecules in the membrane than there are proteins. The membrane proteins on the cell exterior will often have oligosaccharide (carbohydrate) molecules attached to them, which form a coat on the cell surface.

There are different ways in which membrane proteins can be associated with the phospholipid bilayer. Transmembrane proteins extend through the lipid bilayer and, like the lipids, are amphipathic, consisting of hydrophobic regions which interact with the lipid tails within the membrane, and hydrophilic regions that protrude the membrane and are exposed to water. This morphology is determined by the position and nature of the amino acid side chains. The final tertiary or quaternary structure of the protein will place side chains with polar functional groups to the outside in the regions exposed to water, while greasy side chains that are hydrophobic interact with the lipid membrane.

Other membrane proteins are located within the cytoplasm, attached to the cell membrane by means of covalent bonding with fatty acid chains, while other membrane proteins are entirely exposed to the extracellular medium and are attached to the cell membrane by covalent bonds to specific oligosaccharides. Additional proteins may be bound to these integral membrane proteins by non-covalent interactions and are known as peripheral membrane proteins. These types of proteins have different functions. The transmembrane proteins are often involved in molecular transport across cell membranes. Extrinsic proteins serve as cell signalling receptors.

Transport across plasma membranes is a crucial part of a cell's existence. The cell membrane presents a barrier to most polar molecules, which is important for maintaining concentrations of solutes in the cytoplasm. Likewise, the membrane-bound organelles within the cell can have a specific concentration of molecules contained within: different from that of the cytoplasm or extracellular medium. However, critical substances required by the cell must have a means of entering the cell as well as the removal of waste products. This is where the key role of transmembrane transport protein comes into fruition as they are responsible for transporting these water-soluble molecules across the plasma membrane. A given transport protein will be responsible for assisting the movement of closely related groups of organic molecules, or a specific ion, across the membrane. There are two classes of membrane transport protein: carrier proteins and channel proteins. Carrier proteins have moving parts, activated by the chemical energy source ATP, that mechanically move small molecule across the membrane. This is known as active transport. Channel proteins form a narrow hydrophilic pore that enables the passive movement of inorganic ions, known as facilitated diffusion. By these mechanisms, the cell can create large differences in composition between the internal environment and extracellular medium. This is essential for specialised cells to perform their role in the body.

To elaborate by revisiting previous examples: the pancreatic cells must release insulin into the bloodstream to control blood sugar level; in the mitochondria hydrogen ion transport is required for ATP synthesis; neurons must transport sodium and potassium ions to produce an electrochemical gradient. Transport proteins are instrumental for all these processes. One can acknowledge that malfunctions in these processes are likely to result in disorders. Cystic fibrosis is a disease caused by a defect in a transport protein for chloride ions. Diabetes results from malfunctions in the control mechanisms of homeostasis; Alzheimer's, dementia and other neurological disorders are all challenging diseases at this time and understanding the molecular processes behind the conditions will help in the development of treatments.

4.1.6 CELLULAR SIGNALLING

For a cell to operate normally as a single unit, besides the requirement of transport proteins to allow passage of essential nutrients, intracellular signalling proteins are needed to generate a response to changing internal conditions. In multicellular organisms such as ourselves, cells must also interact with one another. An array of cell surface receptors enable response to many chemical messages simultaneously and sensitively. Extracellular signalling molecules are recognised by specific receptors on the surface membrane, or within the target cells. There are hundreds of signalling molecules with which eukaryotic cells communicate, including proteins, small peptides, amino acids, nucleotides, fatty acid derivatives, steroids, retinoids and even dissolved gasses, such as nitric oxide and carbon monoxide. These communicating agents may be released from the cell by exocytosis, where they are carried in a lipid vesicle, or by simple diffusion through the plasma membrane. Some may remain bound to the cell if their purpose is signalling to local cells. The recipient cells, to which the communication is directed, will have specific receptors that are a complementary fit with the signal molecule and upon binding propagate a cellular response.

There are three forms of signalling mediated by secreted molecules: paracrine, synaptic and endocrine. Signalling molecules secreted by cells act as local mediators, which only effect cells in the immediate environment, must not be allowed to diffuse too far, so are rapidly taken up by the neighbouring cells, or destroyed by extracellular enzymes, or immobilised by the extracellular matrix. This is called paracrine signalling. For multicellular organisms to be able to coordinate cell behaviour across the entire organism, some signalling molecules must travel far afield to distant cells. This is achieved in two ways: by networks of nerve cells and by the action of hormones. Synaptic signalling involves routes of neurons along which electrochemical impulses travel to stimulate the release of chemical signals called neurotransmitter, which carry the signal between neurons across gaps called synaptic junctions and propagate the electrochemical impulse in the adjoining neuron. Endocrine cells release hormones, which are signalling molecules that travel in the bloodstream of an animal (or sap in plants) and thus is distributed widely throughout the body, enabling signals to be carried to distant cells. Since this process relies on diffusion, it is much slower than synaptic signalling.

Autocrine signalling involves cells secreting signalling molecules that can bind to its own receptors and coordinate decisions between groups of identical cells. This generates the 'community effect' observed between cells and enhances the effect of signalling when groups of cells carry out the response simultaneously. Activity can also be coordinated between cells through gap junctions, which can form between cells with closely apposed plasma membranes, where the cytoplasm of neighbouring cells is joined directly by narrow water-filled channels. This enables the exchange of small intracellular signalling molecules, such as Ca^{2+} and cyclic AMP; therefore, gap junctions allow cells to communicate directly unrestricted by the barrier presented by intervening plasma membranes.

Cell signalling processes allow the cells that make up multicellular organism to communicate with one another, be it their neighbours or distant cells elsewhere in the body. Any given cell may be exposed to hundreds of signals, not all of them relevant to its particular function. In order to perform their specific role in the body, each cell is programmed to respond to specific combinations of signalling molecules. Furthermore, different cells respond differently to the same chemical signals. This means that the behaviour of cells can be controlled in highly specific ways. These signalling molecules generate a response by binding to cell surface receptor proteins, which act as signal transducers, where the binding of the signalling ligand induces an intracellular response that alters the behaviour of the target cell. There are three known classes of cell surface receptor, defined by the mechanism of transduction used. Ion-channel-linked receptors are those involved in the rapid synaptic signalling between neurons during an action potential (nerve impulse), mediated by neurotransmitters. G-protein-linked receptors indirectly regulate the activities of target proteins bound separately to the plasma membrane, for example, enzymes or channel proteins. The mediator for this process is called 'G protein', which binds to the receptor, and then upon binding of the signalling molecule to the receptor, becomes activated, and then departs to bind to, and activate, the target enzyme or channel protein. The third type of receptors are enzyme-linked receptors, which become activated by a signalling molecule and prompted into acting as an enzyme.

Chemical messages are critical for coordinating cellular activities and maintaining normal function of the body. They are responsible for stimulating the body's organs to operate, prompt the growth and repair of tissues and maintain homeostasis. In order to carry out these processes, cells must manufacture the appropriate raw materials, such as proteins and nucleic acids, which are needed for DNA replication, and the enzymes required for cell division. Malfunctions at any stage of these sensitive biochemical processes can lead to disease. Understanding the molecular basis of diseases aids medicinal chemists in designing molecules that have the required properties needed to target the erroneous receptor, protein, gene: whatever may be the source of the disease and interact with the target in a beneficial way to mitigate the symptoms of the disease and serve as a therapy. Medicines are made by finding a target and designing a drug that will influence the cellular activity associated with the target in a way that can restore good health.

4.2   THE MOLECULAR BASIS OF DISEASE

The design of a medicinally useful compound is a long and arduous process. The first difficulty is identifying a biological target for which a therapeutic molecule can be designed that will interact in such a way as to combat the disease. Understanding the molecular basis of disease is paramount in order to be able to design a compound with the correct chemical structure to not only bind to the biological target but interact in a way that produces a biological response that suppresses the malfunctions caused by the disease that have an adverse effect on good health. Once a target is identified and a target molecule determined through computational analysis, organic chemists strive to synthesise this lead compound. Many slightly differing structural analogues of the lead compound may need to be produced to optimise the properties of the drug. This includes minimising side effects. After a series of clinical trials, and billions of dollars of investment, the drug may become available to market.

The work undertaken by medicinal chemists has been instrumental in the improvements observed in the health of society and the increase in life expectancy in modern times. Surgical procedures that are now routine, prior to the development of antibiotics, carried great risk of septicaemia and death. Many diseases caused by harmful pathogens can now be treated. Malfunctions of the body or mind that are understood on a molecular basis can also be cured by medicines. Conditions that were once life-threatening, such as diabetes or heart disease, now have drugs that are available to preserve a healthy life. As people are now living longer, society is faced with the challenges of an aging population. Diseases such as cancer and neurological deterioration in Alzheimer's and dementia are now starting to be understood and drugs are available on the market which strive to combat these diseases. It is important to acknowledge how much chemistry has improved our lives.

4.2.1 TARGETING THE CELL

Drugs operate in the cell by interrupting a biochemical process such that there is a change in the activity of the cell, causing a noticeable effect. They may enter the cell via transport proteins in the cell membrane, or if they are suitably hydrophobic pass straight through the phospholipid bilayer. On the other hand, the drug might target receptors on the outer surface of the cell. How a drug reaches the target is an important factor of consideration for the medicinal chemist. The drug molecules must reach the site of action in sufficient concentration to be effective and this may affect the route of administration of the compound. Compounds that are not assimilated well when taken orally may have to be injected. Chemists can alter the properties of the molecule to make it more orally bioavailable, but this must not impede the interaction of the drug with the target. Drugs can be designed to target specific enzymes, cell receptors and even nucleic acids. A drug molecule's ability to interact with the target is paramount to determining its efficacy.

Drugs that target enzymes are designed to inhibit their normal operation. This can be achieved in different ways. **Competitive inhibitors** mimic the molecular structure of normal substrate so that the drug can bind with a complementary fit to the active site of the enzyme. This has the effect of blocking the active site, preventing entry of the normal substrate. The necessary reaction with the normal substrate cannot proceed, hence the biological process is subdued. The extent of this effect depends on the concentration of the drug, which in turn determines how many active sites are inhibited out of the plethora of catalytically available enzymes. Also important is the strength of binding of the drug to the active site, which affects the length of time that the drug remains in the active site, impacting the probability of normal biological catalysis happening. The nature of non-covalent interactions, being changeably broken and re-formed, results in inhibition occurring dynamically. The weaker the intermolecular forces between the drug and the active site, the greater the proportion of unencumbered enzymes at any one time and the biological process will be inhibited to a lesser extent. The drug must be designed to optimise non-covalent interactions in order to be effective. Alternatively, the drug molecule could perhaps be designed to undergo reaction once in the active site to form a covalent bond to an amino acid residue. This is an irreversible form of inhibition and renders that enzyme molecule redundant.

Not all enzyme inhibitors operate within the active site. Another method is **non-competitive inhibition**, where drugs that bind to a non-functional part of the enzyme surface can be designed, and this physical binding may cause distortion of the protein structure, having the effect of changing the shape of the active site so that the substrate will no longer undergo complementary fit with the enzyme. Hence, the normal biological function is interrupted.

Enzyme inhibitors have been used widely in medicine. To combat infections from microorganisms, shutting down enzymes that are crucial to the function of the bacterial cell will kill the cell or prevent proliferation. It is possible to selectively target bacterial enzymes without effecting our own due to the large biochemical differences between bacteria and ourselves. For example, some of the first antibiotics were the sulphonamides. These acted as competitive inhibitors and medicinal chemists synthesised a library of these compounds to optimise binding interactions to improve efficacy. These drugs were the antibiotics of choice prior to being superseded by penicillin, which also functions as competitive enzyme inhibitors, except on a different target. In the fight against viruses, successful antiviral drugs have been developed that work against viral enzymes. Acyclovir for the treatment of herpes and saquinavir for HIV are both enzyme inhibitors. Besides battling against foreign invaders, enzyme inhibitors can be utilised to work against the body's own enzymes and thus regulate and offer control over the cell's biological operations. Anticholinesterases are an example where inhibitors of enzymes were developed for control of problems with the nervous system.

Inhibition of enzymes interrupts cell biochemistry and subdues the biological process for which that enzyme is responsible. Many of these biological processes are activated by cell signalling processes. Operations of the cell can therefore be interrupted from the start by targeting the receptors for these chemical messages.

In order to maximise intermolecular binding interactions, the receptor protein changes conformation to accommodate the chemical message. This change of shape in the receptor induces a cellular response to the signal. It is therefore clear that to design a drug that targets a specific receptor, it must satisfy these non-covalent interactions and partake in complementary binding. Essentially, the drug must mimic the natural substrate in an analogous way to enzyme–substrate complex formation. Moreover, given that cellular receptors control biochemical processes within the cell, drugs can be designed that either enhance or suppress cellular activities. Drugs that are designed to closely mimic the natural substrate are **agonists**; they have the effect of increasing the concentration of chemical messages so that the receptor is more frequently activated and the biochemical process to which that receptor is linked will be enhanced. In some instances, a drug may be required to suppress the activity of the cell: **antagonists** are designed to mimic the natural substrate closely enough to interact with the receptor but are sufficiently different as to not trigger a response upon binding. This has the effect of decreasing the availability of receptors and suppress the activity.

To design an agonist, the drug must have the correct binding groups placed in the right positions for binding, and indeed be the correct size for the binding site of the receptor. The binding groups are those that form non-covalent bonds with the receptor and may comprise polar functional groups, producing electrostatic interactions, or greasy hydrocarbon sections that will form hydrophobic interactions. When designing a drug, these groups will be placed at precise positions of the target molecule to maximise binding forces with the receptor. To achieve this, a medicinal chemist must have an understanding of how the messaging molecules interact with the receptor. This is known as **pharmacodynamics**. Typically, this can be done using X-ray crystal diffraction studies of the isolated target protein to elucidate the nature of the protein's binding site. From this information, binding interactions with the drug can be inferred and tailored to control activity of the drug inside the organism. This is referred to as **pharmacokinetics**.

When a drug is required to suppress the activity of a receptor, binding interactions must be designed so that the drug behaves as an antagonist. Molecular modelling and X-ray crystallography can be used to reveal the structure of the binding site, which enables deduction of drug pharmacodynamics so that binding groups can be positioned on the target molecule to produce a drug that binds to the receptor, but in such a way that the change in conformation of the protein receptor is incorrect for propagating a response to the signal. Generating a library of target molecules, each with slightly different binding properties, to find a potentially successful drug, the suite of molecules can be screened against the binding site to optimise the pharmacokinetics and yield a final medicinally useful compound.

There are also alternative ways that antagonists can be designed, without complementary binding to the receptor site. In an analogous method to non-competitive enzyme inhibition, allosteric antagonists may bind to a completely different part of the receptor protein, resulting in distortion of the receptor site, which is therefore unable to accommodate the natural substrate. Alternatively, binding can take place with amino acid residues outside of the receptor site and the steric bulk of the drug molecule may block entry to the receptor binding site and prevent access of the natural substrate. This is referred to as the 'umbrella effect'. It is a useful strategy because it may be difficult to develop a drug to target the receptor binding site if molecular modelling of this region is unclear.

Described so far, drugs can be designed to interact with enzymes and protein receptors to control specific biochemical processes and have a medicinally useful effect. The habitual operations of the cell are controlled ultimately by the genetic code inscribed on the cell's DNA, which resides within the nucleus. Drugs can be designed to target DNA, which has huge potential for medicinal therapies. Some of the most notable drugs that target DNA are used in cancer treatment. There are different ways that these drugs act on the DNA molecule and are classified accordingly, including intercalating agents, alkylating agents and chain cutters. In therapies for genetic illnesses, which arise from abnormalities in the patient's DNA, molecular biology and genetic engineering have produced rapid advances in the understanding of genetic diseases, such as haemophilia, cystic fibrosis and many others. Many examples of drugs designed to target enzymes are involved in the battle against pathogens: microorganisms that cause infectious diseases. This has, in the past, proved a successful strategy to combat infectious diseases.

TUTORIAL QUESTIONS

1. Describe how the structure of nucleic acids relates to their functions in the cell for controlling the flow of genetic information and gene expression. Include the following in your answer: Provide a detailed description of the structure of DNA. Explain how non-covalent interactions confer the stability of DNA. Comment on the structures of the nitrogenous bases and the influence this has on hydrogen bond formation.
2. Outline four types of RNA and describe their functions in the cell.
3. Using your knowledge about mesomeric effects, explain why proteins have defined secondary structures and how this effects the structure and bonding involved in the formation of the tertiary structure.

5 Drug Design, Synthesis and Development

ABSTRACT

The interactions of a drug in the body are referred to as pharmacodynamics and pharmacokinetics. Drugs are designed to optimise binding interactions with their target. This may not necessarily require the strongest possible binding, as the effects of the drug may need to be reversible, and this will also have implications for the dosing regime of the drug. The pharmacokinetics of how a drug travels needs to be considered for the entirety of the drug's passage through the body through absorption, distribution, metabolism, excretion and toxicity, abbreviated as ADMET. The binding interactions that are required of the drug need to be carefully considered during the synthesis of the drug. Binding groups need to be positioned on the correct place for optimised binding interactions. Therefore, aspects of the regioselectivity of the reactions used to make the drug must be carefully selected.

Drugs are oftentimes designed to interact with a biological target. The interactions of the drug with its target need to be carefully optimised to ensure the desired medical result is achieved. When designing new drugs, medicinal chemists must take into consideration both the interactions of the drug molecule with its target and the compound's behaviour in the body. **Pharmacodynamics** examines how to optimise a drug's binding to its target. **Pharmacokinetics** considers how a drug travels through the body to reach the target. A compound that has the best binding interaction is not necessarily the most medicinally useful as efficacy depends on the drug reaching the target at therapeutic concentrations. Often, a compromise between pharmacodynamics and pharmacokinetics has to be made.

Basic principles of pharmacodynamics, the branch of pharmacology that is concerned with the effects of drugs and the mechanism of their action, are to understand the physiological and biochemical effects that drugs have and to correlate these effects with the chemical structure of the compound, considerations of the chemical energetics and rates of reaction need to be understood.

Considerations for the pharmacodynamic properties of drugs have been discussed in previous chapters; optimising drug binding with the target is an essential feature of drug design. Designing molecular features to facilitate a drug's journey to the target is equally important. There are five main aspects of pharmacokinetics to take into consideration when designing drugs: absorption, distribution, metabolism, excretion and toxicity, abbreviated as ADMET.

5.1 IDENTIFYING A DRUG TARGET

There are several different approaches that are taken to design a drug against an identified target. Possible targets may include nucleic acids, cell receptors or enzymes.

DOI: 10.1201/9781003516835-5

IUPAC defines a **pharmacophore** as 'an ensemble of steric and electronic features that is necessary to ensure the optimal supramolecular interactions with a specific biological target and to trigger (or block) its biological response'. It is, in essence, an abstract description of the molecular features that are necessary for molecular recognition of a ligand by the receptor site of a biological macromolecule. Structurally diverse ligands, so long as the pharmacophore is present, may be able to bind to a common receptor site and have their specific bonding interaction. When screening a compound library of potential 'hit compounds' ligands with optimal bonding strength can be identified and selected for further medicinal research and development.

The pharmacophore points of test compounds are those that are involved in forming bonding interactions. Pharmacophore features include hydrophobic centroids, aromatic rings, hydrogen bond acceptors and donors and anions and cations. These features must match the reciprocal structures in the receptor site. This leads to a **structure-based drug design**.

The necessary binding groups to optimise bonding interactions with the target receptor must be in the correct spatial position to satisfy the geometry of the chemical bonds and directionality on all non-covalent interactions. This is determined via pharmacophore modelling approaches to drug design. As such, an appropriate molecular scaffold needs to be developed that holds these binding groups in precisely the right position.

It may then become necessary to use bioisosterism to reduce toxicity, improve bioavailability or otherwise improve the efficacy of the drug. **Bioisosteres** are chemical substituents or groups with similar physical or chemical properties which produce generally similar biological responses in the same chemical compound. Switching one bioisostere for another during the drug design process can have the result of slightly changing the biological effect that the chemical compounds have.

The antibacterial chalcones in Figure 5.1 show how small alterations to the generically similar chemical structure can be achieved to enhance the pharmacological effects of the compound. The chalcone structure in Figure 5.1(b) has low solubility and has a minimum inhibitory concentration of 20 µM, whereas the other two chalcone structures, Figure 5.1a and c, have a high solubility and reduced minimum inhibitory concentration by making small alterations to the key bioisosteres.

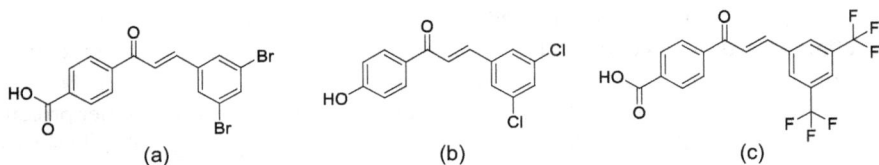

FIGURE 5.1 Antibacterial chalcones as an example of variable bioisostere permitted for enhanced pharmacological effects.

5.2 PHARMACOKINETICS AND PHARMACODYNAMICS

Drug absorption refers to the route or mechanism by which a drug reaches the blood supply, which is also dependent on how the drug is administered. The preferred method of administration is the oral route because it is the least intrusive, hence is used most commonly. When taken orally, the drug is delivered to the gastrointestinal tract (GIT), first entering the stomach, where it is subjected to gastric juices and hydrochloric acid. These chemicals, used to digest food, will also degrade drug molecules, so for a drug to be effective orally, it must be tolerant to these conditions.

In order to enter the bloodstream, the drug must pass the cells lining the intestine. Once the drug has traversed the cell membranes, it can readily enter capillaries through pores that exist between the cells of the blood vessels. Upon entering the capillaries, the drug is en route to the liver, which contains enzymes designed to intercept and modify foreign chemicals in order to make them easier to excrete. This process is known as drug metabolism. This introduces the necessity for an orally active drug to be metabolically stable as well as resistant to the enzymes and hydrochloric acid of the GIT, and this presents a challenge to medicinal chemists designing new drugs.

In addition to metabolic stability, a drug must have the correct balance of water solubility versus fat solubility. Hydrophilic drug molecules that are too polar may fail to pass through the fatty cell membranes of the gut wall. A drug that is non-polar could be too hydrophobic and poorly soluble and dissolute into globules in the gut. In both these instances, absorption would be poor. To help balance the dual requirements of water and lipid solubility, the amine function group is often incorporated into drug molecules, and amines can often also be involved in target binding. Amines act as weak bases, so are partially ionised at blood pH and readily equilibrate between their ionised and non-ionised forms, which means that they have good water solubility, but will also pass cell membranes as the non-ionised form.

Typically, the hydrophilic/hydrophobic character of a drug determines how readily it will be absorbed. The size of the molecule also has an influence on drug absorption. As well as physical constraints, larger molecules are likely to contain more polar functional groups, so in principle are less readily absorbed through cell membranes. Certain criteria have been postulated for an orally active drug to be effective, known as Lipinski's rule of five: so-called because the numbers involve multiples of five. The drug's molecular weight must be less than 500 g mol^{-1}, it must have no more than five hydrogen bond donor groups, no more than ten hydrogen bond acceptor groups and a log P value less than 5 (measure of hydrophobicity).

Polar groups that break these conditions tend to be poorly absorbed and need to be administered by injection. Alternatively, drugs can be designed to hijack carrier proteins. If the drug bears a structural resemblance to some of the polar biomolecules used in the cell's natural biosynthetic pathways, the appropriate carrier protein embedded in the cell membrane can be utilised to transport the drug molecule across the cell membrane and into the cell. For example, fluorouracil is transported by carrier proteins for the nucleic acid bases thymine and uracil. Also, small polar drug molecules with molecular weights less than 200 g mol^{-1} can pass through small pores between the cells lining the gut wall, or they can be carried in vesicles (pinocytosis).

Drug distribution is the next step in the pharmacokinetics of a drug. Once the drug has diffused into the capillaries, it is speedily distributed around the bloodstream, and then more slowly distributed to various tissues and organs. Drug molecules do not have to cross cell membranes in order to leave the blood system and enter tissues. Capillaries contain small pores between cells through which drug-sized molecules can pass, but not important plasma proteins. If the drug target is an extracellular receptor, the drug can find the target, bind to it and perform its function. Drugs that have targets within the cell, such as enzyme inhibiters, or drugs acting on nucleic acids must leave the extracellular aqueous fluid and traverse cell membranes to access their targets. To do this, the drug must be sufficiently hydrophobic, or able to utilise carrier proteins or be taken in by pinocytosis. If a drug is excessively hydrophobic, it can become absorbed into fatty tissue, thus lowering the concentration available to bind to the target. Likewise, drugs that are ionised may be less available in the blood supply if they become bound to macromolecules; for example, they can become irreversibly bound to blood plasma proteins.

Designing a drug that can pass the blood–brain barrier is a challenge. Capillaries that supply the brain are lined with tightly fitting cells that do not contain pores. Furthermore, they are coated with a fatty layer formed from neighbouring cells, which presents an additional barrier to the passage of drugs. Consequently, polar drugs, for example, penicillin, cannot enter the brain. However, this has the advantage of designing drugs to be more polar to avoid CNS side effects. The ability to cross the blood–brain barrier is important for the activity of opiates. Polar groups must be temporarily masked using the prodrug approach, or design compounds that utilise pinocytosis in order for the compounds to reach the site of action.

The extent of drug metabolism is another contributing factor to the pharmacokinetic properties of a drug. Drugs may be reactive with a range of metabolic enzymes, which aim to modify the foreign molecule into a structure, known as the metabolite, which is more readily excreted. These transformations may lead to loss of activity of the parent drug, or the metabolite may even have a different activity and lead to side effects or toxicity. Hence medicinal chemists must design drugs consciously not to produce unacceptable metabolites. Indeed, it is a requirement that all metabolites of a drug are characterised before it is approved, including stereochemistry and tested biologically. On the other hand, metabolic pathways can be important for prodrug approaches.

The body has systems in place to remove foreign chemicals. Polar molecules can be quickly extracted by the kidneys, but non-polar molecules must undergo metabolic transformations to make them more polar before they can be excreted. Non-specific enzymes, such as cytochrome P450 enzyme in the liver, can add polar functional groups to a wide range of compounds to improve their water solubility and make then more easily excreted through the kidneys in urine. A different series of enzymatic reactions might reveal polar functional groups that are masked. For example, there are enzymes that can demethylate a methyl ester to unveil a more polar hydroxyl group. This metabolite can be excreted more efficiently.

These reactions are classed as phase 1 reactions and generally involve oxidation, reduction and hydrolysis, typically taking place in the liver. These reactions also tend to involve characteristic functional groups. Some of the structures most prone to oxidation are N-methyl groups, aromatic rings and the terminal positions of alkyl chains. Reduction with reductase enzymes will occur with nitro, azo and carbonyl groups, while amides and esters are prone to hydrolysis by esterase enzymes. Knowledge of the metabolic reactions that are possible for different functional groups enables medicinal chemists to predict the metabolite that a drug is likely to produce. A drug may undergo several of these reactions and yield many different metabolites. Understanding this chemistry helps with drug metabolism studies to identify exactly which metabolites are formed and hence explicate the future prospects of the drug in regard to safety.

Stereochemistry is an important consideration when designing drugs, as chirality can have important implications for drug metabolism. Metabolic enzymes can often distinguish between two enantiomers, such that each can undergo different reactions. This means that testing each enantiomer separately during screening is necessary and may mean that the design of a drug may need an asymmetric synthesis.

Within the liver, another series of reactions, phase 2 reactions, proceed and regularly involve conjugation reactions, where a polar molecule is attached to the drug at a polar functionality, which may have been placed there by a previous phase 1 reaction. The resulting conjugate will be more polar, thus more readily excreted in the urine or bile.

Metabolic stability is an important property of a drug, given that metabolism usually results in loss of activity, or even result in toxicity. This also has implications for dosing. Another issue is that the activity of metabolic enzymes varies between individuals, particularly for cytochrome P450 enzymes. The difference in the activity profile of these enzymes between patients will affect how fast a drug is metabolised and therefore the dose of the drug must be monitored, depending on the rate of metabolism. Indeed, different countries can have different recommended dosing levels of a drug because of the differences in metabolism between populations. Pharmacogenomics is concerned with screening different populations: how genetic variation affects an individual's response to a drug, and therefore gauge the appropriate dosing level.

Metabolism by cytochrome P450 enzymes is also affected by other chemicals. For example, other drugs and even certain foods can influence these processes. For certain drugs, Brussels sprouts and cigarette smoke can enhance activity, whereas grapefruit juice suppresses it. Because a person's diet can influence how a drug is metabolised, recommendations are usually given regarding what foodstuff should be eaten while taking the medication. For example, the immunosuppressant drug cyclosporine has improved activity if taken with grapefruit juice, being less speedily metabolised, but if the antihistamine terfenadine, which is a prodrug, is taken with grapefruit juice, inhibition of metabolism that produces the active compound means that terfenadine persists in the body and can have cardiovascular side effects; hence the active ingredient is now administered directly and is marketed as Allegra.

Drug–drug interactions, where one drug affects the activity of another, can arise when certain medicines are taken together. A common scenario is when a person takes antibiotics while on other medication. Many antibiotics inhibit cytochrome P450 and this can have consequences for any other drugs being taken. As a result, new drugs are tested to examine whether cytochrome P450 is inhibited or activated because of these challenges.

Drugs that live long enough in the body to perform their function but are metabolised quick enough so as not to cause side effects need to be designed. When taken orally, drugs must pass directly to the liver where a percentage will be metabolised before the drug continues to circulate in the bloodstream. This is known as the first pass effect. This can affect whether or not a drug reaches the target at therapeutic concentration, hence sometimes drugs must be administered intravenously.

Drug excretion from the body can occur in several different ways. Drugs can be diverted from the liver to the intestines in bile and excreted in faeces. Volatile drugs can be exhaled from the lungs, and up to 10%–15% of a drug can be lost through the skin in sweat. Predominantly, drugs are metabolised and excreted via the kidneys. The function of the kidneys is to filter the blood of waste chemicals, which are subsequently removed in the urine. Blood enters the kidneys via the renal artery, which divides into many capillaries, each of which forms a knotted structure called a glomerulus that fits into the opening of a duct called a nephron. Blood entering the glomeruli is under pressure, causing plasma to be forced out through the pores in the capillary walls and into the nephron, along with the dissolved drug and metabolites contained within. Components that are too big to pass through the pores, such as plasma proteins and red blood cells, remain in the capillaries within the remaining plasma. This is a filtration process, so whether the drug is polar or hydrophobic doesn't affect the efficiency of passage into the nephron. However, not all drugs and metabolites will be excreted with equal efficiency, depending on the hydrophobicity of the compound. Some of the compounds that have passed into the nephron will be reabsorbed into the rich network of blood vessels surrounding the nephron before reaching the bladder. As water from the filtered plasma is reabsorbed, drugs and other compounds become concentrated in the nephron. This produces a concentration gradient and hydrophobic compounds may diffuse across the cell membranes of the blood vessels and become reabsorbed in the blood. Polar compounds are retained and dissolve in the urine to be excreted.

5.3 DRUG DOSING

The pharmacokinetics of a drug, which can be broken down categorically as ADMET, are the major determining factors that govern drug dosing. There are many pharmacokinetic variables involved in drug distribution; many are dependent on the route of administration, so estimating the correct dosing regimen can be a challenge, where considerations for the concentration of the drug in each dose and the frequency of administration should be taken into account. Drugs are taken at regular time intervals to maintain a fairly constant level in the blood: not too low as to be ineffective and not so high as to be toxic. This equilibrium is known as the therapeutic dosing level.

Figure 5.2 shows how the therapeutic dosing level is established. Dosing regimen A quickly reaches the therapeutic level, but continues to rise at a steady state to reach a toxic level. Regime B involves half the amount of drug at the same dosing frequency and takes longer to reach the therapeutic level, but the steady state remains below the toxic level, within the therapeutic window.

Complications in drug dosing arise due to differences in age, sex, race, etc. between different patients. For example, people who are obese present a particular challenge because it is hard to predict how much of the drug will be stored in fat tissue. The precise time of drug dosing is another aspect that is difficult to predict

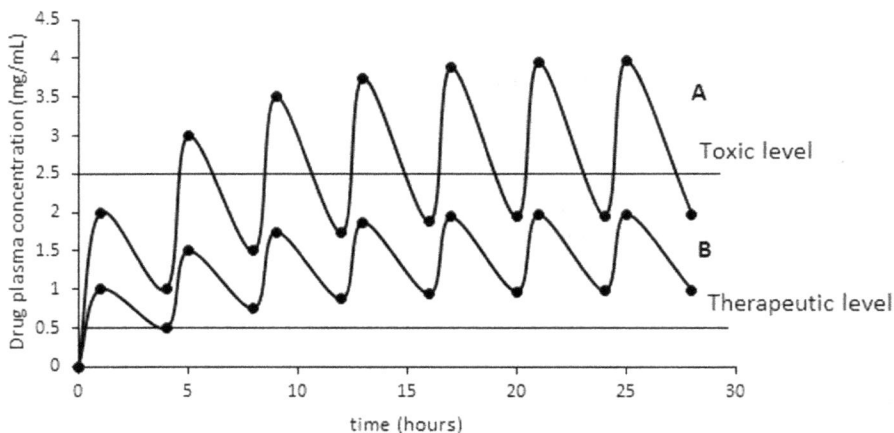

FIGURE 5.2 Two drug dosing regimens showing the change in drug concentration in plasma to illustrate the therapeutic dosing window.

because metabolic reactions proceed at different rates throughout the day; medicines to be taken before bed, when metabolism is slower during sleep, will probably have a slower activity than medicines that need to be taken multiple times throughout the day. One crucial variable for determining how frequently drug dosing is needed is the drug's half-life ($t_{1/2}$), which is the time taken for the blood concentration of the drug to fall by half. This depends on excretion and drug metabolism processes which do not occur linearly with time. Half-lives may vary from a timescale of just a few minutes, such as some of the opiate analgesics, to a number of days, as with diazepam (Valium), where it can take over a week to recover from the effects.

In order to maintain a steady-state concentration in the blood, the dosing must be complementary with the rate of excretion and metabolism, thus it is important to know the half-life of the drug. Generally, the time taken for a drug to reach its steady-state concentration is about six times its half-life. For example, in Figure 5.2, the drug has a half-life of 4 hours and is dosed at 4-hour intervals. Notice half the level is at a maximum after each dose and falls to a minimum before the next dose is provided. It is important to ensure that the blood concentration does not fall below the therapeutic level, or exceed the toxic level; in other words, it should remain in the therapeutic window. The plasma level of the drug at steady state depends on the size of the dose given as well as the frequency; therefore, during clinical trials, blood samples are regularly taken from patients in order to establish safe dosing levels.

Drug tolerance is another aspect of consideration. With some drugs, the effects may diminish with repeat dosing, therefore dosing needs to be increased. This may arise with an increase in production of metabolic enzymes in response to the drug, or the target may adapt to the presence of the drug, for example, antagonism of a receptor may lead to biosynthesis of more cellular receptors, hence more of the drug would be needed. These effects can lead to chemical dependency, where the patient requires the drug in order to feel normal; otherwise, withdrawal symptoms ensue. This relates to drug tolerance in that, if more cellular receptors are produced, the cells become oversensitive to the substrate, be it neurotransmitter or hormone, and it can take weeks for cellular mechanisms to break down the excess receptors, during which time the patient will experience discomfort from withdrawal symptoms. Tolerance to drugs can also lead to increased risk of side effects and overdose, and the danger of the user incurring harm is far greater.

5.4   HOW DRUGS ARE MADE

Medicinal chemists strive to make compounds that serve as remedies for poor health. In order to achieve this, a thorough understanding of the cellular and molecular biology of the ailment is required as well as an extensive knowledge of synthetic organic chemistry. First, a viable target must be identified for the disease; next identification of a hit compound, which can be achieved by molecular modelling of hit–target binding interactions, and screening a range of related compounds in the lab, or computationally to find the compound with the optimum structure–activity relationship. Lead optimisation is the next phase of drug development, where the binding properties of the hit are further optimised through modification of the molecular structure to maximise drug–target interactions, but also taking into account the pharmacokinetic properties of the drug.

Organic synthesis is of paramount importance for creating new drugs. Knowledge of the chemistry of different functional groups enables medicinal chemists to select particular chemical reactions to use to combine molecular fragments and build a predetermined molecular structure of a drug molecule, based on the outcome of computational analysis and lead optimisation processes. Similarly, the required functional groups that are needed for binding to the target can be introduced to the molecular structure by selecting the appropriate chemical reactions for adding a new functional group, or transforming an existing functional group into the one that is required. The chemical reactions needed to produce the novel compound are set out chronologically to yield a recipe for making the new drug, and this is referred to as a synthetic strategy.

The molecular framework of drug compounds commonly comprises aromatic or heteroaromatic ring structures. These can serve as a scaffold for holding the functional groups involved in binding interactions in the correct positions, and can be involved in binding as well. Aromatic and heteroaromatic compounds have a diverse chemistry and offer many opportunities for chemical reactions to introduce new functional groups, or amalgamate molecular fragments, needed to build a target compound.

Benzene commonly reacts with electrophiles via electrophilic substitution reactions. For example, nitration, sulphonation and halogenation reactions, which all require a Lewis acid catalyst, an acceptor of electrons, are useful ways of introducing new functional groups to aromatic rings. Another useful example for synthesis would be the Friedel–Craft reactions (acylation and alkylation) because these form new carbon–carbon bonds, useful for combining molecular fragments.

Reactions of substituted benzenes depend on the functional groups already present on the ring. Regioselectivity, which governs where on the ring the reaction takes place, depends on the electronic effects of the substituents and their position on the ring. Activating groups are typically +I and +M groups, where positive induction and positive mesomeric effects donate electrons into the aromatic ring system. Deactivating groups withdraw electrons from the ring (−I and −M) and destabilise the ring, making it less reactive. Activating groups direct the reaction to the ortho and para position (depending on steric factors of the substituted benzene), while deactivating groups direct to the meta position, except for halogens, which are o/p directors. During synthesis, it is therefore important to add the substituents in the correct order to achieve the desired regioselectivity.

Heteroaromatic compounds are more reactive than benzene and will readily react with electrophiles at the C2/C5 position, due to more resonance structures being available to stabilise the positive charge of the transition state. Six-membered hetero-aromatics, such as pyridine, will also undergo nucleophilic substitution at C2 and C4 because the nitrogen acquires a negative charge through resonance. Methyl pyridines don't behave in the same way and instead react at the methyl group due to the −I effect from nitrogen. These compounds therefore offer interesting alternative ways of assembling a target compound.

When it comes to building a drug molecule, carbonyl chemistry offers a rich variety of reactions that can be used to create new functional groups, or make new carbon–carbon bonds, which are so important for combining different molecular fragments together that comprise the final target compound. Ketones and aldehydes are versatile functional groups and will undergo a diversity of reactions. The electro-negative oxygen polarises the C=O bond, making the C susceptible to nucleophiles and the αH easily deprotonated with a base. Predominantly, ketones and aldehydes will undergo nucleophilic addition reactions; this includes reduction with hydrides to an alcohol and addition of nucleophiles to the carbonyl carbon. A range of nucleo-philes can be used, including alcohols, amines and cyanide, which are particularly useful in synthesis to introduce new functional groups, and Grignard's reagent to make a new C–C bond. All of these basic organic reactions of carbonyls are invalu-able for building a target compound.

The α hydrogens of a ketone can readily be deprotonated with a base: α substi-tution reactions involve replacement of the αH with another group via an enolate. Useful reactions in synthesis include aldol condensations, where two ketones are combined in an alkylation reaction. In each case a new C2–C3 bond is formed. This is a common strategy in organic synthesis because of the versatility of the C2–C3 bond in making new compounds. Ketones can also be used as a starting material for making alkenes, which provides an opportunity for a whole new set of organic reac-tions and alternative synthetic routes towards producing a target molecule.

Often it may not be feasible to introduce the required functional groups straight away. This could be due to the types of reactions being implemented in the synthesis and reaction selectivity for different functional groups. As a result, reactions known as functional group interconversions may be required towards the end of a synthetic strategy to unveil required functional groups. Reduction of organic functional groups is a common reaction used for this purpose.

In many cases, the stereochemistry of the drug molecule is of critical importance. Consequently, the synthetic strategy needs to account for stereochemistry in regard to the staring materials being used and the stereoselectivity of the chemical reactions utilised in the synthetic strategy. Often an asymmetric synthesis is required, which will involve specific reagents and chemical processes.

The synthetic strategy for making a drug can be devised by working backwards from a given target molecule. This approach is called retrosynthetic analysis. In this method, organic syntheses can be planned by breaking down a target molecule into smaller building blocks that are commercially available, known as readily available

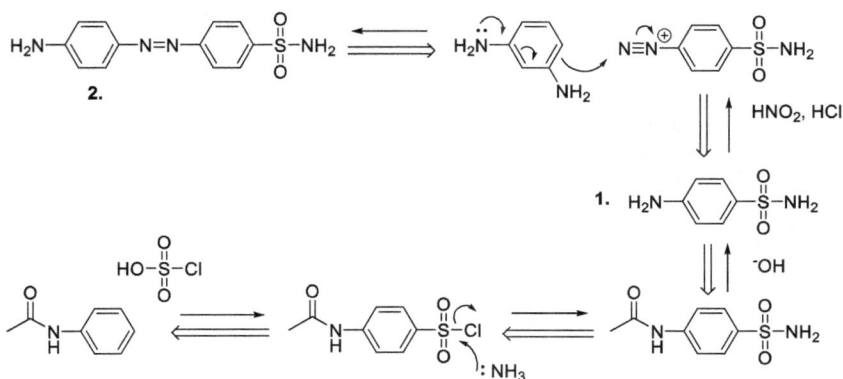

FIGURE 5.3 RSA for the antimicrobial drug prontosil red.

starting materials. This is done by the imaginary breaking of bonds, referred to as disconnections, indicated by an open-ended arrow to produce synthons, or by functional group interconversions. Synthetic equivalents are then used to accomplish the forward reaction in a synthesis.

An example RSA is given in Figure 5.3. Bayer started to explore azo dyes as antibiotics and this led to the development of sulpha drugs, such as prontosil red (**2.**). Despite this drug being effective *in vivo*, it did not affect bacterial cultures *in vitro*, suggesting that an active pharmacophore is produced in the body, later discovered to be structure **1**. Consequently, further synthesis from **1.** was no longer necessary. These compounds can be easily made, and a range of analogues were synthesised to find the most effective, and at the same time reveal the structure–activity relationships of this class of drugs.

The synthesis starts from benzamide, where the amide group acts as an o/p director and directs the reaction with the sulpha group on to the 4-position of the ring. Chlorine is then substituted by ammonia, followed by base-catalysed cleavage of the carbonyl group to produce the active compound. A functional group interconversion is implemented, transforming the amine to the azo group. Electrophilic aromatic substitution reaction, supported by the electron-donating properties of benzamine, produces the azo dye. This synthesis was used over 100 years ago during the preparation of azo dyes and it was later found that the compound had medicinally useful properties. However, these days, more modern synthetic techniques are used during the manufacture of drug molecules.

The current most widely used reactions include amide bond formation, Suzuki–Miyaura coupling and nucleophilic aromatic substitution (not electrophilic aromatic substitution), which are often Pd-catalysed, enabling the use of milder reaction conditions. Figures 5.4 and 5.5 show different reaction schemes that represent the use of modern synthetic techniques.

The Suzuki–Miyaura coupling is frequently used for the synthesis of biphenyl moieties in organic compounds. Figure 5.4 contains two known drugs with biphenyl motifs made via Suzuki–Miyaura technology. Structure (A) is that of losartan, and angiotensin II receptor antagonist, synthesised in 1995, and structure (B) is a more modern example, synthesised in 2014, a HCV NS5A inhibitor, ledipasvir.

(a)   (b)

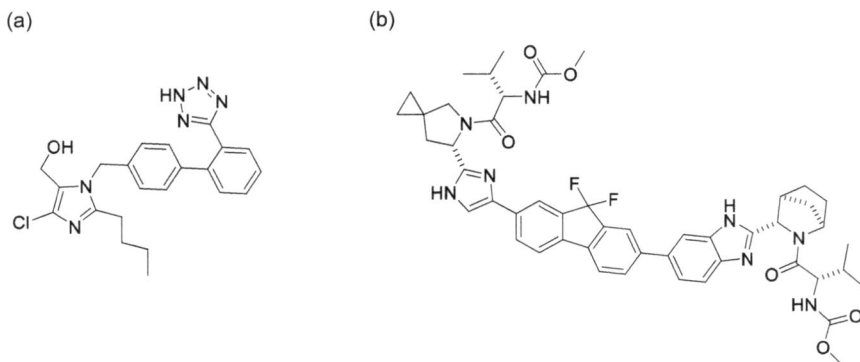

FIGURE 5.4 Examples of drug compounds synthesised using Suzuki–Miyaura type coupling for the creation of biphenyl moieties in organic compounds.

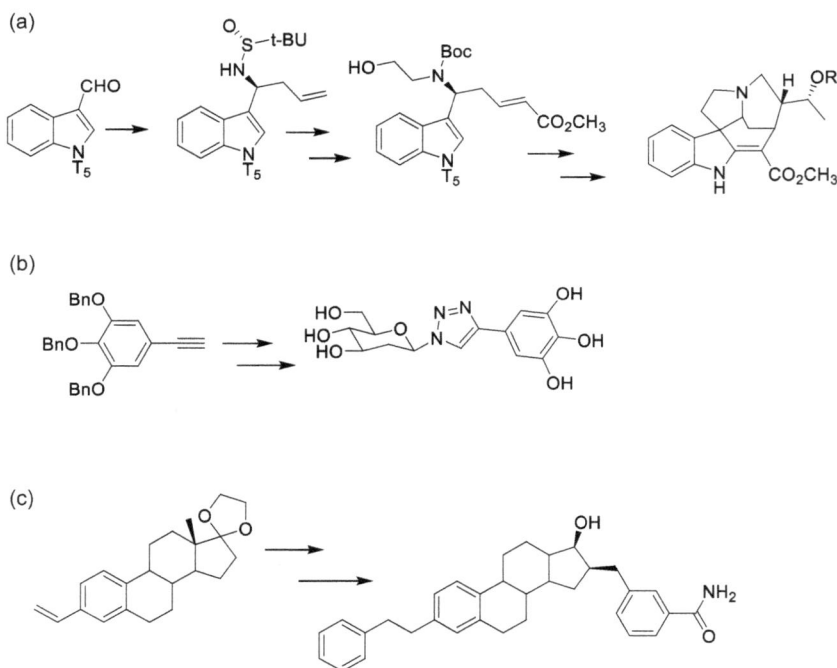

(a)

(b)

(c)

FIGURE 5.5 Examples of modern synthetic techniques in the summarised reaction schemes for the production of medicinally useful compounds.

The reaction scheme, Figure 5.5(a), uses chiral amino allylation, followed by cross metathesis reactions. The reaction in Figure 5.5(b) uses Click chemistry. Boron-based compounds can often be used as cross-coupling partners. Cross metathesis reactions are used during the reactions in Figure 5.5(c).

5.4.1 PREPARATION OF DRUG COMPOUNDS

Once the synthesis of a drug is complete, the compound is not yet ready for consumption. Thorough purification steps are required to clean the organic compound. Chromatography is a commonly adopted technique for separating a target compound from the rest of the matrix. In chromatography, the column is packed with a solid stationary phase consisting of very fine particles and a solvent is selected, referred to as the mobile phase, in which the target compound dissolves. The desired target product will have a different affinity for the stationary phase than any by-products or unreacted material, and therefore will have different retention times on the column. The product elutes from the column at a different time to the contaminants, so can be isolated. The greater the length of the column, the greater the degree of separation, as in the case of high-performance liquid chromatography (HPLC) and gas chromatography (GC), used for volatile compounds, and even separation of stereoisomers is possible.

Chromatography can be coupled to mass spectrometry (MS), which enables the relative molecular mass of the analytes to be accurately determined and helps to characterise product and contaminants. However, this will not give information on isomers. Spectroscopic techniques are required for complete characterisation. Infra-red spectroscopy is a quick and useful technique for identifying functional groups that are present in the molecule, but does not yield complete structural information and this is where nuclear magnetic resonance (NMR) spectroscopy is needed. This enables accurate structural characterisation, and information from coupling constants and integration can determine the proportion of each isomer present.

Understanding organic reaction mechanisms and the step-by-step processes by which molecules can be combined and altered enables chemists to design novel compounds. This is of fundamental importance to drug discovery because new drug molecules can be rationally designed to interact with a biological target associated with a disease and therefore have a biological effect that improves the condition of the patient. Subtle modification of molecular structures and lead compound optimisation can be done by implementing techniques in organic synthesis to yield the best possible drug properties of a lead compound. A key example of where this is important is asymmetric synthesis, where a stereospecific drug–target interaction is required, which could otherwise lead to side effects; note the case of thalidomide. While it is imperative to synthesise drugs with the least possible side effects, it is also essential to ensure the purity of drugs. This is the role of the analytical chemist, who has the important occupation of screening for and eliminating contaminants. Analytical techniques have improved greatly in modern times; contaminants can be traced below nanogram levels, and methods such as spectroscopic techniques can be used to characterise the contaminants. As technologies advance in the future, medicinal chemists will be well equipped to manage the challenges that are presented to the field of medicine in the coming decades.

TUTORIAL QUESTIONS

1. Enzymes can be useful drug targets. Explain how drugs can be designed to target enzymes.
2. Briefly explain the significance of Lipinski's 'Rule of Five'.

6 Pathogens and Immunity

ABSTRACT

Diseases caused by pathogenic microorganisms historically were a great challenge, as scientists did not have a firm understanding of infectious disease, or the mechanisms by which they manifest. However, remedies for infections have been available throughout history, sourced from natural products. Although scientists did not understand how certain natural products remedied infections, they recognised that an active component was effective against the disease. As medicine advanced, these natural active compounds could be isolated and characterised and even modified to become more effective. The discovery and subsequent development of penicillin is a good example. Today, infectious diseases continue to be a challenge for medicine, with the remarkable adaptability of microorganism, they gain the ability to become resistant to antibiotics, and viruses mutate so rapidly that drugs that once worked well become redundant. As such, continual development of new lines of drugs to battle infection is required and this may require looking beyond natural products and involve the rational design of novel compounds against an identified target.

Our fight against pathogenic microorganisms is as old as humanity itself. We have evolved to cope with pathogens by having an immune system that is exquisitely well adapted to recognising foreign invaders and eliminating them. However, some pathogens prove harder to overcome. Historically, humans have found remedies in nature that can assist the immune system and help cure the ailment. With the advent of modern medicine, these natural products have been characterised and isolated for use in medicine. For example, penicillin was discovered in mould spores. Latterly, the structures of natural products can be modified by medicinal chemists to prove even more effective at treating infections. Furthermore, medicinal chemists can even identify molecular targets that they can rationally design drug compounds for and thereby take a strategic approach to making new medicines, as was pioneered during the development of antiviral agents, such as those to combat HIV.

6.1 ANTIBIOTICS: THE NEED FOR INNOVATION

In the past, bacterial infections had a high likelihood of resulting in morbidity as the ailed person relied on their body's immune system to overcome the infection. Very few remedies were available that could alleviate the infection; these mostly included medicinal herbs and dubious tonics, which have been used throughout history. Prior to the development of antibiotics, infection from pathogenic bacteria was a major cause of death. The development of antibiotics in the 20th century led to a dramatic increase in life expectancy. From a cultural point of view, these drugs are perhaps among the most valuable developments in 20th-century medicine.

Bacterial infections continue to be an important health issue, particularly in developing countries, and as resistant strains develop, medicinal chemists must

DOI: 10.1201/9781003516835-6

continue to develop new strategies to 'battle the bugs'. A total of 1.3 million people died from tuberculosis (TB) in 2022. In 2022, an estimated 10.6 million people fell ill with TB worldwide (World Health Organization (WHO), 2022). However, the incidences of TB globally are falling by 2% each year, and thanks to effective diagnosis and treatment, an estimated 54 million lives were saved between 2000 and 2017. However, multidrug-resistant TB (MDR-TB) remains a public health crisis and a risk to security, emphasising the necessity for new front-line drugs to be developed. The outbreak of antibiotic-resistant pathogens is an ever-present challenge to medicine, and novel compounds must continually be developed to tackle these germs. Infections that cannot be readily treated can lead to sepsis, which is the clinical manifestation of acquired infections, which arises when the body's response to the infection injures its own tissues and organs. It is thought to affect tens of millions of people, potentially resulting in millions of deaths, according to studies reviewed by the WHO. Many diseases caused by bacteria are prominent in developing countries, where medical infrastructure is limited. For example, cholera outbreaks result in the loss of tens of thousands of lives, even though many cases can be treated effectively with oral rehydration therapy (ORS), if not antibiotics. Pneumonia accounts for 16% of deaths of children. Pneumonia caused by bacteria can be treated with antibiotics, but only one-third of children that fall ill of the disease receive the antibiotics they need. *Staphylococcus* infections are a relevant problem, even in developed countries, particularly with the upsurge of MRSA. This reflects the need for new drug development.

Historical uses of antibacterial herbs or potions have been documented in several countries, including the Chinese using mouldy soybean curd to treat boils and wounds. Greek physicians used wine for treating wounds, myrrh as well as inorganic salts and honey was used on wounds in the Middle Ages. Of course, these people did not understand that infections were caused by microorganisms. Indeed, bacteria were only first identified in the 1670s by Leeuwenhoek while using microscopes to observe microscopic entities in water; it was not until the 19th century that their link with disease was realised thanks to experiments by Pasteur.

Edinburgh surgeon Josef Lister was an early advocate of a 'germ theory of disease' and used carbolic acid as an antiseptic and to sterilise the operating theatre to avoid infecting his patients. This improved surgical survival rates drastically. Later in the 19th century, scientists such as Koch were able to identify the microorganisms responsible for a particular disease. Paul Ehrlich was a pioneer in finding chemical agents that could interfere with the proliferation of microorganisms, while using concentrations that were tolerable by the patient, the so-called magic bullet strategy. This toxicity can be represented by a therapeutic index. By 1910, Ehrlich had successfully developed the first fully synthetic antimicrobial drug. This was the arsenic-containing compound salvarsan, which is particularly noted for the treatment of syphilis and was used until 1945 when it was replaced by penicillin.

However, salvarsan was only effective against a small number of bacteria; a much broader antibiotic was required. In 1935 it was found that a red dye called prontosil was effective against streptococcal infections in the blood and this led to the development of a family of broad-spectrum antibiotics known as the sulpha drugs, or sulphonamides, which were the only effective treatment against systemic infections until penicillin became available in the 1940s.

Despite been discovered in 1928, it was not until 1941 that effective methods of isolating penicillin were developed by Florey and Chain. This drug revolutionised the battle against infection. However, penicillin is not effective against all types of infection. Since it was discovered that penicillin is a toxic fungal metabolite that kills bacteria and allows the fungi to compete for nutrients, it encouraged scientists to investigate microbial cultures from across the globe in search of other possible therapeutic agents. In 1944, the systematic search of soil microbes revealed the antibiotic streptomycin, which extended the range of therapy to the tubercle bacillus and a variety of gram-negative bacteria. Continued research led to the discovery of the other major classes of antibiotics: peptide antibiotics, tetracyclin antibiotics, macrolides, cyclic peptides; synthetic agents including cephalosporin C, isoniazid, nalidixic acid, ciprofloxacin and many others.

Thanks to antibiotics, a wide range of diseases have been brought under control, including syphilis, tuberculosis, typhoid, bubonic plague, leprosy, diphtheria, gas gangrene, tetanus and gonorrhoea. Antibiotics represent a great achievement for medicinal chemistry when considering the hazards that society faced in the period before penicillin. Within living memory, mothers risked septicaemia in childbirth, ear infections in children leading to deafness were common, pneumonia was a frequent cause of death in hospitals, tuberculosis was a major problem and minor injuries could lead to severe infection, requiring amputation of a limb, and the risk of peritonitis lowered the success rate of surgical operations. Perhaps we take antibiotics for granted today; they are readily available and used frequently for mild conditions such as a chesty cough, and few of us appreciate the profound effect that these drugs have had on quality of life.

The success of antibiotics relies on the drugs being damaging to bacteria and acting selectively against the bacterial cell wall and not affecting the cells of the patient. This can be achieved through understanding the bacterial cell, as they tend to have different structures and biosynthetic pathways to animal cells. Bacterial cells are classified as prokaryotic and have many differences to eukaryotic animal cells. Bacterial cells do not have a defined nucleus to contain their DNA like animal cells do. Instead, the DNA is held loosely in the cell, or on a ring structure, called a plasmid. Animal cells have many other organelles besides the nucleus, such as endoplasmic reticulum, mitochondria among others, whereas bacterial cells are relatively simple regarding their internal structures. The different constructs between prokaryotic and eukaryotic cells relate to significant differences in biochemistry. For example, bacteria must synthesise many of the crucial molecules they require, such as essential vitamins, which animals obtain from food. Targeting these metabolic pathways unique to bacterial cells is one strategy for designing drugs to combat infections. Antimetabolite drugs selectively target bacteria, not affecting the host, by inhibiting enzyme-catalysed biochemical pathways that occur in the bacterial cell, but crucially not in the animal cells. Another mechanism by which antibiotics can be designed to combat bacteria is to directly target protein synthesis to prevent the enzymes and other proteins that are essential to the bacteria from being made. Inhibition of nucleic acid transcription and replication is another way to disrupt protein synthesis and disrupt cell division.

Sulpha drugs are an example of antibiotics that affect bacterial cell metabolism. It was discovered in 1935 that a red dye called prontosil red was effective against bacteria *in vivo* (i.e. in laboratory animals), but antibacterial effects were not observed *in vitro*; cultures in petri dishes were unaffected. It was later discovered that bacteria present in the small intestine of the animal metabolised prontosil to give the true antibiotic product called sulphanilamide. Prontosil was an early example of a prodrug.

Sulphonamides proved effective against a wide range of infections and further developments extended the range to a number of gram-positive bacteria, especially *pneumococci* and *meningococci*. There were limitations to sulphonamides; they proved ineffective against *salmonella* and typhoid, and problems arose in how the drugs were metabolised, often producing toxic by-products. Hence sulpha drugs were eventually superseded by penicillin. Prior to the development of penicillin though, sulpha drugs were the antibiotics of choice for infectious diseases and are noted for saving Winston Churchill's life after falling ill during the Second World War. Sulphonamides have been particularly useful against intestinal infections. Succinyl sulfathiazole is a prodrug of sulfathiazole; the succinyl moiety contains an acidic group, which becomes ionised in the weakly alkaline conditions in the intestine. As a result, it is not absorbed into the bloodstream and is retained in the intestine. Slow enzymatic hydrolysis of the succinyl group releases the active sulfathiazole where it is needed. Examples are given in Figure 6.1.

Structure–activity relationships were deduced from the synthesis of a large number of sulphonamide analogues. The para-amino group must be unsubstituted because of their function; the only exception is where an acyl group is used, which is inactive, but is metabolised in the body to generate the active compound, i.e. amides serve as a prodrug. The aromatic ring and sulphonamide functional group are both required. The sulphonamide nitrogen must be primary or secondary and is the only site that can be varied.

A compound library of sulphonamide analogues are often varied by introducing heterocyclic or aromatic structures to the sulphonamide nitrogen, which affects the extent to which the drug binds to plasma proteins, hence controls the blood

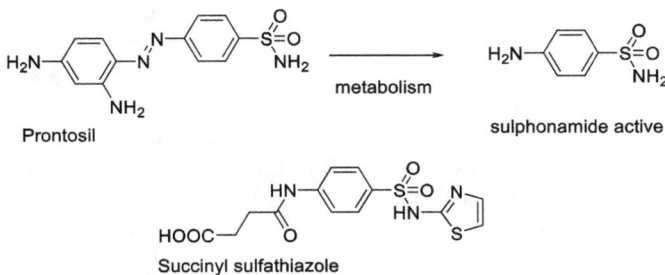

FIGURE 6.1 A reaction scheme showing the metabolism of prontosil red to form the sulphonamide active (above) and succinyl sulfathiazole (below), an example of a sulphonamide prodrug.

levels and lifetime of the drug. A drug that binds strongly to a plasma protein will be more slowly released into the blood circulation and will be more long-lasting. Variations of SNHR, where R can be different ring structures for example, also alter solubility, affecting the pharmacokinetics of the drug, rather than its mechanism of action. For example, an N-acetyl group makes the compound insoluble, causing it to clog the kidneys and therefore is toxic. The compounds must contain an N–H proton, which has a low pKa value, meaning that it is readily ionised, and the drug becomes soluble.

The mechanism of action of sulphonamides is to work as competitive enzyme inhibitors and block the biosynthesis of folate in bacteria. Folate is an essential nutrient for bacteria, required for DNA replication. With restricted synthesis of folate, growth of bacterial populations subsides, and the immune system can handle the infection. Humans acquire folic acid in food; bacteria are unable to do this and must synthesise their own, so there are no side effects to human health from deficiency of this nutrient. This kind of target specificity is referred to as the 'magic bullet' approach. By structurally mimicking the normal substrate for this enzyme, sulphonamides competitively inhibit it because the active site is tricked into accepting the drug, which therefore blocks access to the active site, albeit reversibly, to the substrate, and reduces the production of folate.

Resistance can be acquired by bacteria that produce more folate than usual, thus competition for the active site favours the normal substrate. Alternatively, resistance can be acquired through mutations that alter the structure of the active site subtly to reduce the affinity for sulphonamides over the natural substrate. Mutations can also arise that decrease the cell membrane permeability to drugs, and this is another issue regarding antibiotic resistance.

One very important difference between bacterial cells and those of animals is the presence of a cell wall that encapsulates the bacterial cell. This is in addition to a plasma membrane and serves to protect the cell, give it structure and stability, but most crucially, to balance osmotic pressure; otherwise if bacteria without a cell membrane entered an aqueous environment containing a low concentration of salts, water would freely enter the cell due to osmotic pressure, which would cause the cell to swell and eventually burst. This is known as lysis. While the cell wall does not prevent water from entering the cell, it does stop the cell from swelling, therefore indirectly preventing excessive amounts of water entering the cell, which would be catastrophic for the bacterium.

Bacteria can be classified according to their cell wall by using a staining technique. A purple dye is added, followed by washing with acetone. Bacteria with a thick cell wall (20–40 nm) absorb the dye and are stained purple. These are defined as gram-positive bacteria. Bacteria with a thin cell wall (2–7 nm) only absorb a small amount of dye, which is washed out with acetone. These bacteria are then stained pink with a second dye and are defined as gram-negative bacteria. Although they have a thin cell wall, the important difference between these bacteria and gram-positive bacteria is that they have an outer cell membrane made up of lipopolysaccharides. This difference has important implications for the different vulnerabilities of gram-positive and gram-negative bacteria to antibiotics.

6.1.1 Targeting Bacterial Cell Walls

In 1877 Pasteur and Joubert discovered that certain moulds produced substances that are toxic to bacteria, but some of these were toxic to humans also, and therefore had no apparent application in medicine. In 1928, Fleming noted that a bacterial culture which had been exposed to the air for several weeks had become infected by a fungal colony. The area surrounding the fungal colony where the bacterial colonies were dying was correctly attributed to the fungal colony producing an antibacterial agent. Realising the importance of this, he cultured the mould, which was found to be a relatively rare species of *penicillium*, a spore of which must have originated in another laboratory and by good fortune blown on the wind into Fleming's laboratory. By chance the weather was just right for this to happen: a cold period allowed the fungus to grow while the bacterial population remained static; then a warm spell enabled the bacteria to grow and the antibiotic properties of penicillin to be observed. As a final twist, the culture plate was also stacked in a bowl of disinfectant above the surface of the sanitised water, ready to be washed up!

Through Fleming's keen observation, he was able to make this discovery and spent several years studying the novel antibacterial agent. He was able to demonstrate that this compound was non-toxic to mammals but was not able to isolate and purify the unstable compound. This problem was solved by Florey and Chain in 1938, who used chromatography, and by 1941 were ready to conduct the first clinical trials. The resounding success led to the commercialisation of penicillin and by 1944 there was enough of it available to treat casualties from the D-day landings.

By this time, penicillin was in widespread use, but despite its popularity, chemists still debated its molecular structure, with many unusual characterisations being held in much scepticism. Surprisingly, it turned out that the structure of the compound was unusual. Dorothy Hodgkin established the structure by X-ray crystallographic analysis and revealed the β-lactam ring. The atoms in the ring are held tightly in a square structure, which is very strained and susceptible to breaking down, hence Fleming's difficulty in isolating the compound.

Synthesis of such a highly strained compound was highly challenging, but manufacturing the compound without the need to laboriously purify it from mould cultures was advantageous for commercialisation of the antibiotic. This was accomplished by Sheehan in 1957. However, the full synthetic procedure was too involved to be commercially viable, but the following year a biosynthetic intermediate of penicillin, 6-aminopenicillanic acid (6-APA), was isolated by Beecham's and revolutionised the production of semisynthetic penicillin by producing a readily available starting material (Figure 6.2).

The four-membered β-lactam ring is unstable, being highly strained, and is fused to a five-membered thiazolidine ring in a bent shape. Its structure suggests that it is derived from the amino acids valine and cysteine. The nature of the acyl side chain R depends on the components of the fermentation medium. The first mass production of penicillin used corn steep liquor as the fermentation medium, which contains high concentrations of phenyl acetic acid, and gives benzyl penicillin (penicillin G). Fermentation mediums containing phenoxyacetic acid give phenoxymethylpenicillin (penicillin V).

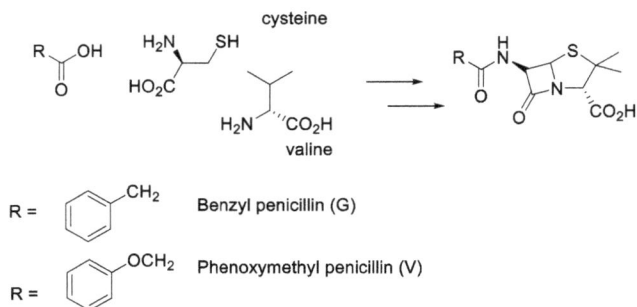

FIGURE 6.2 Overview of the synthesis of penicillin from the amino acids valine and cysteine.

The properties of penicillin G make it a good choice to use as an antibiotic medicine. It is active against non-β-lactamase-producing (i.e. non-resistant) gram-positive bacteria, such as meningitis, gonorrhoea and several gram-negative cocci and anaerobic microorganisms, such as *streptococcus* and *enterococci*, so is active against many rapidly dividing types of bacteria. Penicillin is a non-toxic, very safe drug (magic bullet); however, some people have allergies to penicillin. Allergies range from a mild rash to anaphylactic shock. Small molecules such as penicillin generally do not cause this effect, but nucleophilic groups on proteins will react with the open β-lactam ring and therefore become covalently bonded to the protein, causing an immune response as the protein is 'recognised' as foreign. This only happens in 0.2% of patients. Other limitations to penicillin G are poor activity against gram-negative bacteria, so does not have a broad spectrum of activity. The molecule is broken down by stomach acid, so cannot be taken orally and must be injected, which is the least favourable route of administration to patients. To overcome some of these limitations, it was necessary to synthesise penicillin analogues. These compounds must retain the essential aspects of the molecule crucial to the mechanism of action, while possessing structural modifications that will change the interaction of the drug with the body in such a way as to overcome the limitations of the drug.

To explain the mechanism of action of penicillin, it is necessary to understand the structure of the bacterial cell wall. The cell wall is essential for bacteria to survive in a range of environments, such as varying pH, temperature and osmotic pressure. The cell wall is composed of peptidoglycan; its structure consists of a parallel series of sugar backbones made of two types of sugar, namely N-acetylmuramic acid (NAM) and N-acetylglucosamine (NAG). Peptide chains are bound to the NAM sugars. The presence of D-amino acids in the chain is noteworthy because in humans, biochemistry only involves L-amino acids, which gives the opportunity for target selectivity. Bacteria have racemase enzymes that can convert L-amino acids to D-amino acids, required for biochemical uses such as cell wall synthesis. The cell wall is constructed with the peptide chains linked together by the displacement of D-alanine from one chain by glycine in another. This final cross-linking reaction is the one that is inhibited by penicillin. Consequently, the cell wall framework is no longer interlinked and is fragile; it doesn't stop the cell from swelling, which eventually bursts (lysis) and kills the bacterium. Different types of penicillin are used in this way to inhibit the transpeptidase enzyme responsible for cell wall construction for a range of bacteria. The severity of damage is greater for gram-positive bacteria, which have thicker cell walls, consisting of 50–100 peptidoglycan layers.

The transpeptidase enzyme is classified as a serine protease, where a serine residue in the active site is involved in hydrolysis of peptide bonds. The serine acts as a nucleophile to split the two D-alanine units on a peptide chain. The terminal alanine departs, while the peptide remains in the active site. Another peptide chain enters the active site, and a peptide bond is formed between D-alanine and the terminal glycine of the other chain. It is presumed that the penicillin conformation mimics the transition state conformation of the D-Ala-D-Ala moiety during the cross-linking reaction and the transpeptidase enzyme mistakenly binds it to the active site. The serine residue acts as a nucleophile and opens the β-lactam ring, but because the molecule is cyclic, it is not split in two as the peptide would be. Consequently, nothing leaves the active site, which is blocked and access to the second peptide chain is prevented. As a result, cross-linking in the bacterial cell wall is inhibited, making it fragile and lysis occurs.

Bacterial strains vary in their susceptibility to penicillin, and there are several ways that they can derive resistance. In order to inhibit the transpeptidase enzyme, located on the bacterial cell membrane, penicillin molecules must traverse the cell wall, which is a rigid, but porous structure, so small molecules can readily pass. The cell wall does not present a physical barrier; however, gram-negative bacteria have an outer lipopolysaccharide membrane, which is impervious to polar molecules like penicillin G and water, hence the lower activity of penicillin G against them. Channel proteins called porins facilitate passage of water and essential nutrients across the cell membrane. The structure of this porin, as well as the nature of the penicillin molecule determines the effectiveness of the drug against the bacteria. Typically, small hydrophilic penicillin analogues will more easily pass the outer membrane and be present in high enough concentrations to inhibit cell wall biosynthesis.

Some bacteria develop mechanisms to subvert the actions of penicillin. Mutated forms of transpeptidases result in the presence of β-lactamase enzymes, which open the β-lactam ring to form an ester bond as before, but are able to hydrolyse the ester link and expel the broken-down penicillin. Indeed, some gram-positive bacteria, such as *Staphylococcus aureus*, can release β-lactamase extracellularly and break down penicillin before it even reaches the cell membrane. Currently, 95% of *S. aureus* strains release β-lactamase capable of hydrolysing penicillin G. This presents a huge problem with respect to treating *S. aureus* infections.

Most, if not all, gram-negative bacteria produce β-lactamase, which is trapped between the cell membrane and outer membrane, and so is present at a higher concentration than in gram-positive bacteria, hence the observed lower activity. However, variations in substrate specificity mean that susceptibility to some penicillin types can occur in gram-negative bacteria. Efflux pumps in some gram-negative bacteria are able to remove penicillin from the periplasmic space between the cell and outer membrane. Resistance through mutation can be acquired quickly by bacteria as a consequence of their methods of genetic transfer: fast cell division, called binary fission, and conjugation, where mutated genes are passed directly, which speedily passes genetic information between populations.

With the surge of resistant strains of bacteria, it is essential that medicinal chemists produce a variety of penicillin analogues. Sheehan's full synthetic approach to making penicillin took too long and was too low yielding; therefore, it was not commercially sustainable. This restricted medicinal chemists to fermentation and semisynthetic procedures. A limited number of variations were possible by adding different carboxylic acids to the fermentation process, thus changing the acyl group, but this method is time-consuming. Semisynthetic procedures involved fermentation to yield 6-APA, first isolated by Beecham's in 1959, when it was realised that this biosynthetic intermediate was the same as an intermediate in Sheehan's synthesis, which could be treated with a range of acid chlorides. Chemical methods involving hydrolysis of the acyl side chain is a more efficient method.

To consider the feasibility of a penicillin analogue, it is important to understand the structure–activity relationships of the drug. A strained β-lactam ring is essential. The free carboxylic acid is essential, as the penicillin is usually administered as a salt and the ionised carboxyl group binds to the charged ammonium ion of lysine in the active site. The bicyclic system is important to confer further strain on the β-lactam ring and thereby improve activity. An acylamino side chain is required; sulphur is usual, but not essential and stereochemistry is important. In summary, very little variation is tolerated; modifications are limited to the acylamino side chain. Penicillin analogues needed to be produced that addressed the problems of acid sensitivity, β-lactamase susceptibility and limited breadth of activity.

There are three reasons for the acid sensitivity of penicillin: the strained β-lactam ring readily undergoes acid-catalysed ring opening; the β-lactam carbonyl is highly reactive with nucleophiles, not being in resonance stabilised as with usual tertiary amines because mesomeric effects are impossible due to the increase in bond strain that would arise from a double bond being present within the already strained β-lactam ring (120 degrees ideal vs 90-degree bond angle in the ring). Lastly, neighbouring group participation from the acyl side chain to open up the β-lactam ring serves as a self-destruct mechanism.

The mechanism in Figure 6.3 illustrates how the acyl side chain can be involved in the intramolecular ring-opening reaction. The deprotonated nitrogen atom acts as a nucleophile, attacking the carbonyl carbon with its lone pair of electrons. Electrons from the carbon–oxygen double bond then move to the carbonyl of the β-lactam. Resonance of the electron pair in the β-lactam results in the carbon–nitrogen bond of the ring breaking and nitrogen forms a new bond to the proton that was lost to the solution initially.

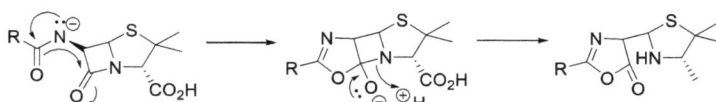

FIGURE 6.3 Mechanism for the intramolecular ring opening of the β-lactam ring, where curly arrows show the movement of electrons around the molecule.

To maintain an effective drug, the only modifications that can be made are to the acyl side chain, which can improve acid tolerance. To reduce neighbouring group participation, an electron withdrawing group is placed in the side chain to draw electrons away from the carbonyl and make it less nucleophilic, for example, the oxygen of penicillin V, therefore this drug can be taken orally because it can survive the stomach acid.

Besides considerations for efficacy regarding the route of administration, modifications to the molecular structure of penicillin become important to combat antibiotic resistance. The need for innovative drug design to produce new antibiotics arises when the bacteria adapt and develop mechanisms that stop the drug from working. Penicillinase/β-lactamase-resistant penicillin needed to be developed when, in the 1960s, widespread use of penicillin G led to an alarming increase in penicillin-resistant *S. aureus* infections.

The strategy of steric shields was implemented; in this approach, a sterically bulky group is placed on the acyl side chain, preventing access of the penicillin to the β-lactamase active site. However, design of a suitably sized steric shield was a challenge because medicinal chemists wanted to avoid compromising complementary fit between the penicillin molecule and the transpeptidase enzyme's active site. Thankfully, methicillin was developed just in time to address the growing number of cases of resistant *S. aureus* infections. The steric shields are the two ortho-methoxy groups on the aromatic ring that are able to make the discrimination between β-lactamase and transpeptidase active sites.

However, there are several drawbacks to methicillin: the absence of an electron-withdrawing group on the acyl side chain means that it is acid sensitive, it has only one-fiftieth the activity of penicillin G against organisms that are sensitive to penicillin G, shows poor activity against some streptococcal strains and is inactive against gram-negative bacteria. As a result, methicillin is no longer used clinically. Better penicillinase-resistant penicillins have since been developed, but, disconcertingly, a large number of *S. aureus* strains that can be detected in hospitals have become resistant to methicillin and other penicillinase-resistant antibiotics as a result of mutation of the transpeptidase enzyme. These are termed methicillin-resistant *Staphylococcus aureus*, or MRSA.

With the perpetual occurrence of antibiotic resistance, the continued need for innovation requires medicinal chemists to work hard to develop new drugs to tackle the upsurge of resist pathogenic microorganisms. Different factors influence the efficacy of a drug and modification to the molecular structure of penicillin may or may not be successful in producing an effective drug. Factors affecting the spectrum of activity of penicillins include: the ability to pass the cell membrane of gram-negative bacteria, susceptibility to β-lactamase enzyme, affinity for transpeptidase target and the rate of removal by efflux pumps. When searching for a novel compound, medicinal chemists must synthesise a library of compounds to find an analogue with the correct structure to overcome these issues.

It is found that hydrophobic groups on the side chain, as in the case of penicillin G, favour activity against gram-positive bacteria, but show poor activity against gram-negative bacteria. Analogues with hydrophilic groups (e.g. NH_2, OH, CO_2H) attached adjacent to C=O (α position) have little effect on gram-positive bacteria, but do enhance activity on gram-negative bacteria, for example, ampicillin and amoxicillin (Figure 6.4).

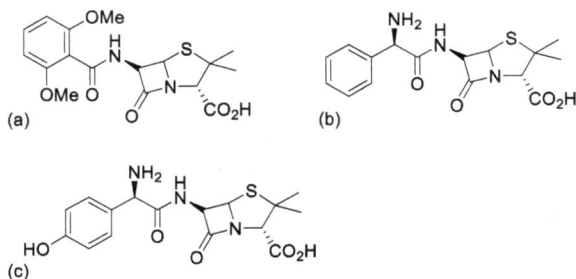

FIGURE 6.4 Molecular structures for the penicillin analogues (a) methicillin, (b) ampicillin, (c) amoxicillin.

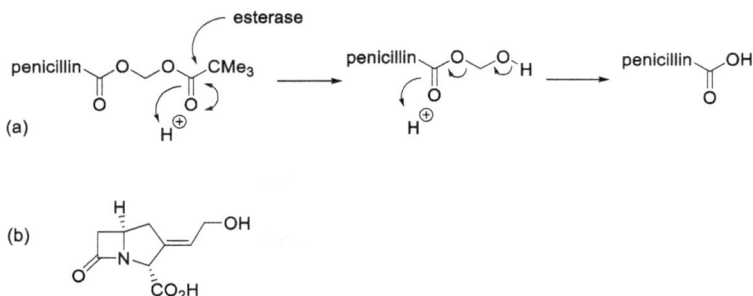

FIGURE 6.5 (a) Curly arrow mechanism to show the removal of the protecting group by esterase enzyme to unveil the active penicillin and (b) molecular structure of clavulanic acid.

The modification of the penicillin by inserting an α-hydrophilic group enhances activity by aiding passage through the porins of gram-negative bacteria's outer cell membrane. These two analogues have a similar structure, but the additional phenol group improves absorption of amoxicillin through the gut wall. The NH_2 group produces acid tolerance, hence these drugs can be taken orally. However, there are limitations; sensitivity to β-lactamase (no steric shield) and they can cause diarrhoea, as poor absorption leads to high concentrations in the gut that can abolish healthy bacterial flora, enabling colonisation of harmful microbes, leading to intestinal distress.

The issue of poor absorption is a result of the dipolar nature of the molecule, which arises from the free amino acid group and carboxylic acid functionalities. The solution is to use a prodrug approach, where one of the polar groups is masked by a protecting group that can be removed metabolically once the prodrug has been absorbed, for example, the use of acyloxymethyl esters, which are susceptible to esterases. A nucleophilic residue in the active site of the esterase cleaves the carbonyl part of the ester from the prodrug, and in a second step the penicillin molecule is unveiled. A curly arrow mechanism for this process is given in Figure 6.5. This can be any analogue mentioned previously, for example, ampicillin protected in this way is called pivampicillin.

For these penicillin analogues, susceptibility to β-lactamase is an issue. However, when used in combination with clavulanic acid, the scope of antibiotics such as amoxicillin is greatly improved. Administered to the patient as Augmentin, the dose level of amoxicillin can be greatly decreased when used in combination with clavulanic acid because clavulanic acid is an irreversible inhibitor of most β-lactamases. It was first isolated from *Streptomyces clavuligerus* by Beecham's in 1976.

Clavulanic acid was the first example of a naturally occurring structure with a β-lactam ring that is not fused to a sulphur-containing ring. Instead, it is fused to an oxazolidine ring. Also note the lack of acylamino side chain. Analogues have since been made, with the essential features for a β-lactamase inhibitor: the strained β-lactam ring, enol ether with Z-configuration, no substitution at C6, (R)-stereochemistry at positions 2 and 5, and a carboxylic acid group.

Drugs containing β-lactam rings are not the only chemicals that inhibit cell wall biosynthesis. Vancomycin is often referred to as the 'antibiotic of last resort' because it interrupts cell wall biosynthesis via a different mode of action to penicillin, so is effective against MRSA. The propensity for bacteria to develop resistance is why vancomycin is restricted to use as a last resort; when doctors have exhausted all other avenues, this is the drug that is used. The hope is that restricting the exposure of bacteria to this drug will limit the risk of adaptation of the microorganisms to acquire resistance and hence prolong the effectiveness of vancomycin. Should bacteria evolve resistance to vancomycin, this would be severely detrimental to the arsenal that doctors have available to treat patients and would promulgate innovation in the field of antibiotics.

Vancomycin is a bacterial glycopeptide produced by a microorganism called *Streptomyces orientalis*. It is derived biosynthetically from a linear heptapeptide chain, containing five aromatic residues, which undergo oxidative coupling with each other to produce three cyclic moieties within the structure (see Figure 6.6). Chlorination, hydroxylation and the final addition of two sugar units complete the structure. The cyclisation transforms a highly flexible heptapeptide molecule into a rigid structure, where the peptide backbone is held in one conformation and hindered bond rotation around the aromatic rings offers further stability.

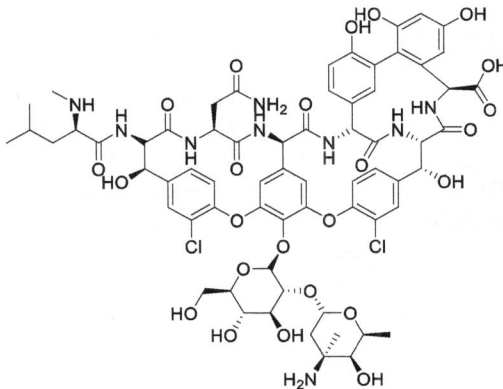

FIGURE 6.6   Molecular structure of vancomycin.

The fixed conformation of the heptapeptide chain is necessary for the unique mode of action of vancomycin. The rigidity is necessary for the formation of five hydrogen bonds to the tail of a crucial pentapeptide cell wall building block. Dimerisation occurs as a second vancomycin molecule forms four hydrogen bonds with the first. The steric bulk of these molecules prevents access to the pentapeptide cell wall component by transglycosylase and transpeptidase enzymes needed for cell wall biosynthesis. Being such a large molecule means that vancomycin cannot pass through the cell membrane, and therefore is not effective against gram-negative bacteria.

Other antibacterial agents are designed to act on the plasma membrane. The polypeptides valinomycin and gramicidin A cause the uncontrolled movement of ions across the cell membrane. Valinomycin acts as an inverted detergent and complexes a 'naked' potassium ion. The hydrophobic outer of the complex can pass the cell membrane and deposit the potassium ion outside the cell, resulting in fatality. Typically, K^+ is in high concentration inside the cell and can only pass the membrane via specialised transport proteins and this equilibrium is disrupted by valinomycin. Valinomycin is selective towards K^+, having the correct spatial arrangement of donor atoms to displace water ligands and form bonds to K^+. Other ions, such as Na^+, are not the correct size to fit and displacing water ligands is too difficult. Unfortunately, the toxicity is not selective to bacteria, affecting mammal cells as well. Gramicidin A is a peptide consisting of 15 amino acids, which coil into a helix where hydrophobic side chains point outwards and interact with the membrane. Two helices of gramicidin A must combine to span the membrane and the hydrophobic interior of the helix serves as a channel for the passage of ions. However, gramicidin A is also toxic to humans. Producing compounds to serve as drugs which employ the 'magic bullet' approach is a challenge for medicinal chemists. Compounds must be toxic to bacteria, but safe to use and to achieve this, target specificity is essential when developing effective antibiotics in the future.

6.1.2 OTHER METHODS TO TARGET BACTERIA

In the battle against pathogenic microorganisms, medicinal chemists have implemented other strategies of drug design besides producing compounds to target the bacterial cell wall. For example, antibacterial agents have been designed which impair protein synthesis. Protein synthesis is orchestrated by the cell's DNA. An enzyme called DNA helicase separates the strands at a specific region on the DNA molecule, the gene containing the instruction for a specific protein, such as those like β-lactamase, which gives rise to resistance. DNA helicase breaks the hydrogen bonds between the DNA bases, enabling another enzyme, called RNA polymerase, to move along the template DNA strand and bind the exposed bases to complementary nucleotides that are present in the cell. This process is known as transcription and results in the production of a stand of RNA which carries a complementary sequence of bases to the template gene on the DNA.

In eukaryotic cells, sections of the gene known as introns do not code for proteins and are removed during a process called splicing. The coding sections, called exons, can be arranged in different combinations, meaning that a single gene can code for several proteins. Once the final copy of the RNA, which is known as messenger

RNA, is made, this is produced directly from transcription in prokaryotes; it is available for translation, whereby the code contained in the sequence of bases is interpreted and the appropriate sequence of amino acids is constructed into a protein.

The sequence of bases on the mRNA are organised as discrete triplet codes; three bases code for one amino acid, and different sequences of triplet bases, each called a codon, code for a particular amino acid. For example, GAC codes for the amino acid aspartic acid. The process of translation is not as straightforward as the amino acids lining up along the mRNA strand. A second type of RNA, called transfer RNA, is involved. The tRNA is a smaller molecule and is responsible for binding free amino acids in the cytoplasm and bringing them to the mRNA template. The tRNA molecules contain an anticodon, which is the opposite sequence to that on the mRNA and is complementary, therefore the tRNA carrying the amino acid can bind to mRNA. Different codons are also present on the mRNA strand to determine where to start and terminate translation into a protein.

Translation is done at organelles called ribosomes. A ribosome attaches to the starting codon on the mRNA molecule. A tRNA molecule, with a complementary anticodon, carries a specific amino acid and attaches to the mRNA by specific base pairing. A second tRNA molecule attaches to the next codon in the sequence in the same way. The ribosome moves along the mRNA molecule, working on two tRNA molecules at a time, and joins together the two amino acids via a peptide bond, using an enzyme and ATP. As the ribosome continues to move along the mRNA stand, the free tRNA molecules break loose and leave to collect another of the same amino acid from the pool of amino acids in the cytoplasm. This process continues until the ribosome reaches a stop codon (one that does not code for an amino acid) and the completed peptide is released. Note that up to 50 ribosomes can pass immediately after the first, so many identical polypeptides can be synthesised simultaneously. It is at this point that the tertiary structure would assemble from its constituent polypeptides.

Selective toxicity against bacteria can be achieved in drugs that target ribosomal RNA, inhibiting different stages of the translation process, due to the fact that prokaryotic ribosomes differ in structure to those found in eukaryotic cells. The bacterial ribosome is a 70S particle, composed of a 30S subunit which binds to mRNA and initiates protein synthesis, and a 50S subunit, which binds to the 30S mRNA complex to make the ribosome. The ribosome has two main binding sites: the peptide site (P site) binds the tRNA bearing the peptide chain, and the acceptor aminoacyl site (A site) binds the tRNA bearing the next amino acid in the protein sequence, to which the peptide chain will be transferred. Eukaryotic cells have bigger ribosomes made of a 60S large subunit and a 40S small subunit.

Streptomycin is an aminoglycoside: a carbohydrate structure which contains a basic amine group, and was the next most important antibiotic after penicillin. It proved to be the first effective agent against tuberculosis. Having been absorbed through the outer membrane of gram-negative bacteria, which was a limiting factor for efficacy of penicillin, streptomycin binds to the 30S subunit of the bacterial ribosome and prevents the triplet code on mRNA from being transcribed, so protein synthesis is terminated and vital proteins are not made.

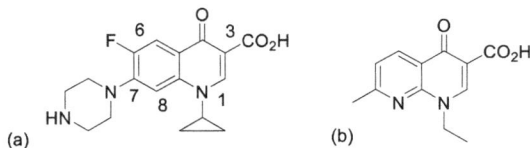

FIGURE 6.7   Molecular structures of (a) ciprofloxacin and (b) its precursor, nalidixic acid.

There is a wide diversity of antibiotics designed to target bacterial protein synthesis. Tetracycline antibiotics are a broad-spectrum class of antibiotics that also bind to the 30S subunit, but have a different mode of action. They prevent aminoacyl tRNA from binding and stop growth of the protein. Other antibiotics target the 50S subunit of ribosomes, such as chloramphenicol, and work by inhibiting the movement of ribosomes along the mRNA strand. Chloramphenicol is the drug of choice for treating typhoid in some parts of the world, where more expensive drugs cannot be afforded.

Other antibiotics target bacterial protein synthesis by acting on nucleic acid transcription and replication. Examples of the structures of this class of antibiotics are given in Figure 6.7. Quinolone compounds, such as nalidixic acid, are particularly useful for the short-term treatment of urinary tract infections. Resistance soon develops to these compounds; consequently, new analogues had to be developed. Modifications to the structure of nalidixic acid were found to increase the spectrum of activity against both gram-negative and gram-positive bacteria.

A single fluorine atom at position 6 greatly increased activity as well as uptake into the bacterial cell. Addition of a piperazine ring on position 7 is beneficial; improved oral adsorption, tissue distribution, metabolic stability as well as improving the level and spectrum of activity are among the advantages. Presumably, the ability for the basic substituent to form a zwitterion with the carboxyl group is the reason for these improved drug properties. Further modifications include addition of an isopropyl ring to nitrogen 1 and replacement of pyridine with benzene. This led to the development of ciprofloxacin, which is regarded as one of the most active broad-spectrum antibiotics available. Furthermore, bacteria are slow to develop resistance to it, unlike nalidixic acid.

6.1.3   THE OCCURRENCE OF ANTIBIOTIC RESISTANCE

The adaptive ability for bacteria to develop resistance is of great concern. A large proportion of *Streptococcus pneumoniae* and *Staphylococcus aureus* strains are resistant to β-lactam antibiotics. More ominously, resistance is even rising against the drug of last resort for treating MRSA, vancomycin. Hence it is imperative that medicinal chemists continue to develop drugs to combat infection. Bacteria can acquire resistance against many antibiotics, and this is known as multidrug resistance. Multidrug resistance can be acquired by the accumulation of multiple genes, each of which codes for resistance against an individual drug, typically on resistance plasmids within the cell. Multidrug resistance can also occur from overexpression of multidrug efflux pumps, enabling many types of antibiotics to be extruded from the bacterial cell. Some gram-negative bacterial strains are resistant to all known antibiotics, notably those belonging to *Pseudomonas aeruginosa* and *Acinetobacter baumanii*, and can serve as a reservoir for transmission of resistant genes. The emergence of so-called 'pan-resistant' bacteria coincides with a regression in novel drug development by major pharmaceutical companies, as the economic incentive to produce antibiotics is low compared to other drugs, such as anticancer agents. Society without these lines of defence to combat infection is a sombre thought.

There are several ways that resistance can develop. Bacterial cells divide and reproduce very rapidly; chance mutations that result in an enzyme or protein that impairs the effectiveness of a drug have a higher probability of occurring the longer the infection persists. Hence it is imperative that a patient completes a course of antibiotic, even if symptoms of the infection subside long before the course is finished. The antibiotic needs to destroy the majority of the bacteria, then the body's immune system can cope with the remaining few that are more resistant. People failing to complete their course of antibiotics is a major factor contributing to upsurge in multidrug resistance.

Once drug resistance through chance mutation has arisen, other bacterial cells can acquire resistance through genetic transfer. Not only is the gene for resistance transferred to daughter cells as resistant bacteria divide, but also genes can be passed between bacterial cells. There are two main ways in which this can occur: conjugation and transduction. In conjugation, the genetic material is transferred directly between bacterial cells, through a connecting bridge of sex pili built between the two cells. Transduction involves small sections of genetic material, called plasmids, being transferred by means of bacterial viruses (bacteriophages) which may leave a resistant cell, then go on to infect a non-resistant cell with the relevant genetic material needed to acquire resistance; by doing so, they pass on the genetic information in the plasmid containing the instruction for resistance enzymes, such as β-lactamase.

Resistance is particularly prevalent in hospitals, where use of antibiotics is greatest. Trace amounts of drug are present in the air in hospitals, and these have been attributed to development of resistance; when breathed in they kill sensitive bacteria in the nostrils and thereby encourage proliferation of resistant strains. The prevalence of resistance is most widely attributed to the overuse of antibiotics. The careless use of antibiotics to treat minor infections in medicine, the widespread use of veterinary medicines including as additives to animal feed, all greatly enhance the possibility of resistance developing. Many bacterial strains are now resistant to the early antibiotics, such as penicillin G, yet in poorer third world countries, where their use is more limited, these drugs are still effective.

The need to develop new drugs is essential to overcome the challenges of bacterial resistance. Sequencing genomes of proteins may elucidate new targets in the ever-present battle against pathogenic bacteria. Many of the compounds produced to combat pathogenic bacteria are semisynthetic, derived from microorganisms and rely on fermentation for manufacturing them. By employing innovative research, targeting on a molecular-level approach might aid in designing novel compounds to battle infection, and overcome one of the leading challenges of medicinal chemistry today.

6.2   ANTIVIRAL AGENTS AND RATIONAL DRUG DESIGN

Viruses must take over a host cell in order to survive and multiply. They are non-cellular infectious agents, essentially a protein packet containing nucleic acid, which is transmitted to the host cell to enable the virus to be replicated. Philosophically, it is debatable whether viruses can be considered as living entities, given that they do not produce their own energy and require living cells for their replication, although a predisposition for self-replication is a property of living things. Whether they may be considered as being corporeal or just lifeless molecular entities, there are a large variety of viruses capable of infecting bacteria, plant and animal cells; indeed over 400 different viruses are known to infect humans.

There are several ways that viruses can be transmitted. For example, airborne viruses, such as influenza, chicken pox, measles, mumps, rubella, viral pneumonia and smallpox can be transmitted in sputum, when an infected person coughs or sneezes. Other viruses require physical contact to be transmitted because they cannot survive long outside a host, for example, the common cold, cold sores (herpes virus), HIV and rabies. Viruses can even be transmitted by arthropods, such as biting ticks in the case of yellow fever. Lastly, foodborne and water borne viruses may be ingested, such as hepatitis A and E and viral gastroenteritis.

6.2.1   THE GLOBAL IMPACT OF VIRUSES

The impact that viruses have on society can be most dramatically demonstrated by historical events. It was speculated that smallpox was responsible for major epidemics that weakened the Roman Empire during the periods AD 165–180 and 251–266. The disease, brought home from travelling soldiers, would have raged for many years, indicated by historical records, killing many people whose immune system was not equipped to handle the disease. Smallpox was also responsible for the decimation of indigenous communities in North and South America, following first contact with Europeans during the early colonisation of the New World. The influenza pandemic in 1918 killed more people in this period than there were casualties in the First World War. It became known as the Spanish flu because of censored press releases, to avoid loss of morale during the war; headlines reported cases in neutral Spain mostly, but illness and mortality in other European countries were rife.

In developing countries, viruses still present a devastating and omnipotent threat. Originally, these outbreaks were contained in small communities, and, although fatal to many in a localised remote area, the risk of a widespread epidemic was smaller. However, with larger populations and the convenience of air travel, rare diseases are much more easily spread around the world. Scientists fear a catastrophic scenario involving the possible evolution of a 'supervirus' with a transmission mode and infection rate akin to influenza, but with a much higher mortality rate.

The global population had a taste of the devastating effects of highly transmissible viruses during the Covid pandemic of 2020. Covid-19, which was thought to have originated in China, spread across the world very quickly on account of its high infection rate and the unpreparedness of the world for a global pandemic. This resulted in unprecedented lockdown restrictions, with the greatest infringement of individual liberties seen in the modern era. This was an attempt to 'flatten the curve' of infection rate to a level that was manageable for the healthcare infrastructure of the host country. This, in turn, led to unforeseen consequences, with excess deaths caused as people were unable to access healthcare as promptly as was needed, resulting in conditions such as cancer diagnosis not being treated; there were also broader problems related to mental health from the isolation of lockdowns and harms to children's education due to school closures. Furthermore, closure of usual commerce meant that the Covid pandemic had a severe economic impact, leading to inflation and a cost-of-living crisis, with the poorest people feeling the effect worst.

However, medicinal chemistry offered solutions. A global effort was directed towards finding a vaccine to cover the population against Covid. The goal of a herd immunity was aimed at so that infection rates could be declined to a manageable level to allow for the natural immunity against Covid-19 to develop. There was, of course, an economic incentive to any company that could develop such a vaccine, so the big pharmaceutical companies pooled their resources into making a special and timely effort to develop a miracle cure. This involved the development of novel technologies in the field of research of antiviral medicine.

While the Covid-19 pandemic claimed many lives, there are viruses that spread rapidly with higher incidences of fatality, but fortunately the latency period between infection and detection of symptoms is short, so the diseases can be contained. However, should a virus evolve that has a latency period that is longer, more people would likely become infected, and this would cause widespread devastation. Since the pandemic of 2020, global communities have feared the occurrence of such a virus, a 'disease X', and global leaders have put into place measures thought to best mitigate against such an occurrence.

With contemplation of historical and recent events, the containment and prevention of viral diseases spreading among populations are of crucial importance to ensuring safety in modern society. Instrumental in managing viral diseases are the medicines developed to combat the infections. Targeting viruses is very difficult, due to the considerable differences between their structure and that of cellular entities. In combatting bacterial infection, antibacterial agents could be found in nature and utilised through fermentation processes to produce antibiotics. Further modifications could be made to the structure to yield novel and effective semisynthetic drugs to combat resistance. Antiviral agents must be designed from scratch based on information from identified targets and this led to the development of rational drug design, a great advancement in medicine. Medicinal chemists now could not only incorporate natural products into designing new drugs but also design and synthesise completely unique chemical structures which interact with biological targets and offer a new line of defence against pathogens.

6.2.2 Structure and Life Cycle of Viruses

In order to be able to identify targets against which drugs can be developed, knowledge of the structure and life cycle of viruses is paramount. A virus particle can simply be considered as a protein package that contains a type of nucleic acid with which it can infect host cells and hijack the host cell's machinery to reproduce itself.

Viruses contain one or more molecules of either RNA or DNA, but not both, and thus are defined as either RNA viruses or DNA viruses. The RNA can be single stranded (ssRNA) as is the case for most viruses, or double stranded, where the base sequence of the RNA that is the same as viral mRNA is called the (+) strand and its complementary partner is the (−) strand. Most DNA viruses contain the typical double-stranded DNA, but single-stranded DNA is present in some viruses. There is great variation between viruses in the size of the nucleic acid, ranging from genomes coding for just three-to-four proteins to larger genomes that code for over 100 proteins.

The protein packet, called the capsid, protects the nucleic acid. The capsid is composed of protein subunits, called protomers, which are manufactured by the host cell and aggregate spontaneously through self-assembly processes. An additional membranous layer may be present around the capsid, containing carbohydrates and lipids. The complete structure is known as the virion and these can range in size from 10 to 400 nm, so can only be viewed by electron microscopy.

Understanding the structure and life cycle of viruses is essential in devising strategies to combat viral infections by identifying suitable targets against which an antiviral drug can be designed. Since the structure of a virion is predicated around its regeneration, finding aspects of the virus's life cycle that can be impeded through intelligent designing of a drug would be hugely advantageous in the development of antiviral agents. Viruses require the host cell's capabilities of protein synthesis to manufacture the constituent parts of the virion, and this is how viruses reproduce themselves. To reiterate, by understanding the mechanisms of this process rational drug design can lead to the synthesis of an agent that will combat the virus.

Viruses display a typical life cycle in which they infect a host cell, hijack the cell machinery to replicate themselves, then leave to infect other cells. The first stage is adsorption; a virion must first bind to the surface of the host cell by interacting with the surface glycoprotein receptors. By doing so, the virus has tricked the cell into accepting the virus as a harmless entity. This leads to the second stage: penetration and uncoating, where the virus introduces its nucleic acid to the cell. This can be done by injecting it through the cell membrane, or by the virus entering the cell, then uncoating. Uncoating occurs in different ways: some viruses fuse with the outer of the cell membrane, then introduce the capsid containing the nucleic acid; other viruses enter the cell via endocytosis, where it is enveloped by the cell membrane, which pinches off to produce a vesicle containing the virion, which is brought into the cell. Enzymes within lysosomes then aid with uncoating.

Once inside the cell, the next stage is replication and transcription of the viral nucleic acid. The virus's genome is incorporated into the host cell genome, then the normal mechanisms for protein synthesis are commandeered; enzymes, ribosomes, amino acids etc. needed to manufacture proteins are utilised by the virus to produce the constituent parts of a virion, and because many proteins can be synthesised simultaneously, enough components are made to produce an inordinate amount of virus particles. These components arrange themselves within the cell to make new capsids through self-assembly. The final stage is virion release. Naked virions, with no outer membranes around the capsid, are released by cell lysis, bursting forth from the cell, leaving destructing in their wake. When the viruses acquire the final components to become mature, they move on to infect other cells.

6.2.3 ANTIGENS, IMMUNITY AND VACCINATION

The body has in place defence systems to combat foreign invaders. The immune system comprises a class of white blood cells known as lymphocytes, which are present in large quantities in the blood and lymph fluid. These cells become activated when they detect foreign invaders, such as bacteria or viruses. Invading microorganisms are detected by lymphocytes from the unfamiliar molecules on the invader's cell surface, known as antigens.

The immune system has the ability to discriminate foreign molecules from 'self' molecules. Nearly any macromolecule that is foreign to the recipient has the potential to elicit an immune response and can be referred to as an antigen (antibody generator). The specificity displayed by the immune system and its ability to differentiate antigens are extraordinary, capable of distinguishing between two proteins that differ by only one amino acid, or even between two stereoisomers of the same molecule.

Immune responses can be broadly classified as either antibody responses or cell-mediated immune responses. Antibody responses involve the production of antibodies by certain cells known as B cells, which circulate in the bloodstream and permeate other body fluids in search of the antigens that induced their creation. Antibodies are proteins known as immunoglobulins and they bind to this specific antigen, which inactivates its toxic effects by subverting the toxin's binding to cell receptors. Antibody binding to antigens on the cell surface of a pathogen marks the invading microorganism, making it readily identifiable by phagocytes; cells of the immune system engulf and destroy the invader. Specialised cells, called T cells, that target the foreign invader, produced during cell-mediated immune responses, react with the antigens on the surface of infected host cells. The T cells may kill the virus-infected host cell, thus eliminating it before the virus has replicated, or in other cases the T cells may produce chemical messages to activate macrophages to destroy the invading virus.

Once the immune system has been subjected to a particular pathogen, it not only adapts to recognise the antigens from that pathogen but also develops a memory, meaning that should this pathogen infect the body again, the immune system will react quickly to combat the infection. The antibodies produced that have high binding specificity to the antigens from this pathogen continually circulate in the blood ever after. The ability of the immune system to remember exposure to particular pathogens is utilised during vaccination.

Vaccination is a preventative approach to protect people from viral diseases and is particularly successful against childhood diseases, such as polio, MMR, smallpox and yellow fever. The approach was first realised by Edward Jenner in the 18th century, when he observed that a milkmaid, who contracted the less virulent disease cowpox, subsequently had become immune to smallpox. He inoculated people with material from cowpox legions and discovered that they too had become immune to smallpox.

Vaccination works by priming the immune system by introducing foreign material that has molecular similarity to some component of the virus, but lacks its infectious nature, or toxic effects. The molecular fingerprint of the virus is recognised as 'non-self' and the cells of the immune system prepare their defence so that, should the virus infect the body, the immune system is ready to attack it. Generally, a weakened or dead virus, or components of the virus that display a characteristic antigen (subunit vaccine) are administered.

However, there are limitations to vaccination. Vaccines usually are not effective on patients that have already become infected by the virus. Also, rapid gene mutation in viruses constantly changes the amino acid composition of glycoproteins on the virus surface, which leads to different antigens that are not recognised by the immune system, so the virus is disguised. Vaccines may not be suitable for patients with a weakened immune system, such as individuals with cancer, who underwent an organ transplant, or with HIV, therefore another strategy is needed: antiviral drugs.

The reason that vaccination generally does not work on patients where infection has already occurred is because, for most of its life cycle, the virus is within a host cell and so is hidden from the immune system and circulating medicines. Moreover, because the virus capitalises on the cell's own biosynthetic mechanisms, the options for potential drug targets are limited compared to other invading pathogens. This makes developing antiviral drugs challenging.

Medicinal chemists are presented with the challenge of devising a strategy to combat viruses and rationalise and design drugs for any potential target. The design of such drugs will be cognisant, based on research into the virus structure and life cycle. Until the 1980s, very few clinically useful drugs were available for treating viral infections. Since then, progress has accelerated due to the AIDS pandemic and improved understanding of viral infectious mechanisms from viral genomic research, which has propagated great advancements in this area. Indeed, a full viral genome can now be quickly determined and compared with those of other viruses, enabling identification of how the virus's genetic sequence is split into genes. While genetic sequence will vary from one virus to another, it is possible to identify similar genes that code for proteins with similar functions, and thus present a possible target.

Genetic engineering methods can be implemented for the production of pure copies of the protein by inserting the viral gene into a bacterial cell, so that sufficient quantities of the protein can be synthesised and isolated to be used for screening. The viability of the protein as a potential drug target can then be determined by studying protein–drug interactions. Proteins that are good drug targets must fulfil certain criteria. They must be crucial for the life cycle of the virus, therefore having a major effect on the mode of infection. The proteins must bear little resemblance to human proteins; therefore, the drug would be expected to have good selectivity and minimal side effects. Ideally, the protein will be common among a variety of viruses with a region of amino acid sequence that is identical and conserved, therefore the drug would have a broad spectrum of activity. Importance for an early stage of the virus life cycle is advantageous, thus the virus is prevented from spreading throughout the body and causing symptoms.

6.2.4 ANTIVIRAL DRUGS IN ACTION

Finding a target to combat viruses is a challenge. A prudent strategy might be to investigate the viral genome and attempt to develop a drug that targets the virus's nucleic acid, given that this is the critical part of the virus needed for its replication.

Different viruses have different nucleic acids, so rationally designed drugs to combat one type of virus may not be effective against another.

Nucleic acids are polymers composed of nucleotide monomers; these building blocks of DNA and RNA are composed of a nitrogenous base (purines or pyrimidines) attached to a sugar molecule and a phosphate group, which make up the backbone of the nucleic acid. The sugar molecule bonded to a base can be referred to separately as a nucleoside.

Most drug development against DNA viruses are targeted against herpesviruses to tackle diseases such as cold sores, genital herpes, chickenpox, shingles, as well as Burkitt's lymphoma and Kaposi's sarcoma. Acyclovir was discovered by compound screening and brought onto the market in 1981. It revolutionised the treatment of these diseases, being the first relatively safe, non-toxic drug to be used systemically for the treatment of these diseases. Acyclovir is a nucleoside analogue, structurally similar to the nucleotides that make up DNA and contains the same nucleic acid base as deoxyguanosine, but lacks the complete sugar ring. The comparison shown in Figure 6.8 illustrates the structural similarities between acyclovir and deoxyguanosine.

Acyclovir is in fact a prodrug; the active agent is generated by phosphorylation in three stages to from a triphosphate within the infected cell. Nucleotide triphosphates are the constituent building blocks of DNA, which assemble along a template strand through the action of enzymes during DNA replication. Namely, DNA polymerase is responsible for this process and acyclovir works as an inhibitor of viral DNA polymerase and prevents DNA replication in two ways: either by inhibiting DNA polymerase or as a chain terminator. DNA polymerase can catalyse the attachment of acyclovir into the growing DNA chain because it has a sufficiently similar structure to deoxyguanosine. Since the sugar ring is incomplete and lacks the required hydroxyl group on position 3′ of the sugar ring, the nucleic acid chain cannot extend further and DNA replication ceases.

DNA replication is essential in uninfected cells also; it is required to replenish healthy tissue, so the risk of side effects needs to be considered carefully. However, selectivity arises as acyclovir is only converted to the active triphosphate in infected cells. Although the enzyme that catalyses phosphorylation is present in normal cells, the herpesvirus has its own version of the enzyme, which catalyses the phosphorylation of acyclovir more rapidly than host cell enzymes. Therefore, in healthy cells, acyclovir remains as the inactive prodrug.

A major limitation of acyclovir is the low bioavailability (15%–30%) when taken orally. Prodrug analogues were developed that remedied this issue by introducing

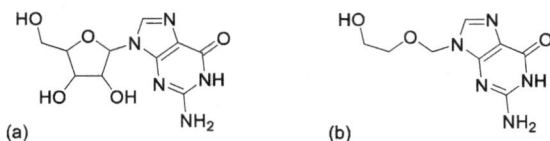

FIGURE 6.8 Structures of the nucleic acid base guanosine (a) and the nucleoside analogue acyclovir (b).

groups to make the compound more water-soluble. Valacyclovir has an L-vinyl ester group that is hydrolysed in the liver or gut wall to unveil acyclovir, which now has a blood level concentration equivalent to intravenous injection.

Exasperatingly, strains of herpes are appearing that show resistance to acyclovir. This is on account of mutations arising in either the enzyme responsible for phosphorylation of acyclovir or in viral DNA polymerase. The consequences of these mutations are that the active drug is not generated in the cell or the active drug is not recognised by viral DNA polymerase and replication of the virus continues uninterrupted.

A very relevant case study to examine drug development against RNA viruses is the treatment of human immunodeficiency virus (HIV). Acquiring this virus causes patients to have a weakened immune system and, without medicine, can lead to acquired immune deficiency syndrome (AIDS) which is usually fatal. Patients with AIDS become fatally susceptible to secondary infections such as pneumonia and fungal infections, which can escalate and lead to death. Scientists discovered that the immune response in patients with AIDS had been weakened by a virus, HIV. The virus infects T cells, which are crucial to the immune system and therefore directly attacks the immune response, meaning that patients are less able to cope with secondary diseases.

HIV belongs to a group of viruses known as retroviruses; there are two variants associated with geographical locations: HIV-1 is responsible for AIDS in American, Europe and Asia, HIV-2 occurs mainly in Western Africa. The extent of prevalence of AIDS has propagated a great deal of research into HIV and currently antiviral drugs have been developed to act against two viral enzyme targets: reverse transcriptase and protease. However, these drugs only mitigate the proliferation of the disease and do not eradicate it; hence ongoing research is directed at finding better targets.

HIV is an RNA virus containing two identical (+)-ssRNA strands in the capsid, along with viral enzymes. The capsid is enclosed by a layer of matrix protein. This virus particle is enveloped by a membrane originating from that of the host cell and contains the viral glycoproteins gp120 and gp41, which are crucial for adsorption and penetration. The gp41 spans the membrane and is bound non-covalently to gp120, which projects from the surface. When the virus approaches the host T cell, the gp120 interacts and binds with a transmembrane protein called CD4 on the surface of the T cell. Conformational change of gp120 unveils gp41 which can now bind to the T cell, and anchors the virus to its host, then conformational change of gp41 pulls the virus towards the T cell so that their membranes can fuse and allow penetration of the HIV capsid.

Once inside the host cell, the protein capsid disintegrates with the assistance of viral protease enzymes. The contents of the capsid are then released into the cell cytoplasm. The viral RNA is not capable of coding for proteins and self-replication directly, but with the assistance of viral reverse transcriptase is converted into viral DNA and incorporated into the host cell DNA. The virus can now remain dormant until cellular processes promote transcription of the viral genes within which the instructions for making the necessary matrix proteins, glycoproteins and crucial viral enzymes are encoded. The viral glycoproteins gp120 and gp41 are produced and incorporated into the cell membrane of the host. As these proteins are manufactured, the constituents of the virion gather together at the membrane, then budding releases the components of the virus. The constituent proteins then self-assemble to produce the virion.

Viral genomics and other research have enabled scientists to gain an understanding of the life cycle of HIV and thus identify suitable targets, namely reverse transcriptase and protease, and develop drugs that interfere with their function. However, the occurrence of mutation is very high in HIV and acquisition of resistance is a big problem. Long term, this can result in selection of mutated viruses that are resistant. As such, using a multidrug approach, where a combination of therapies is administered, tends to be the most efficacious strategy. This method has been successful in delaying the progression to full AIDS but is not a cure.

There are three types of drugs available for highly active antiretroviral therapy (HAART), which may be used in combination: nucleoside reverse transcriptase inhibitors (NRTIs), non-nucleoside reverse transcriptase inhibitors (NNRTIs) and protease inhibitors (PIs). These drugs are largely synthetic in nature, but are based on natural product structures, such as modified arabinose nucleosides, in particular the protease inhibitors which, except for one, are isosteres of the native hexamer peptide substrate. To be efficacious as a therapy, these drugs must fulfil certain requirements. They must have high affinity and selectivity for their targets and thus prevent the virus from multiplying and spreading, while being safe and well tolerated. They must be synergistic with each other so that they can be used in a combination therapy, but also compatible with other drugs that might need to be used to cure secondary infections, such as antibiotics for pneumonia. They should be suitable to take orally, with minimal dosing frequency to maintain therapeutic concentrations in the circulation, because they are likely to be required for the duration of the patient's lifetime. Preferably, the drug will be able to pass the blood–brain barrier in case the virus resides within the brain.

The enzyme reverse transcriptase is a DNA polymerase exclusively associated with the virus, but caution should nevertheless be taken to ensure that there is no inhibitory effect on cellular DNA polymerases when designing nucleoside reverse transcriptase inhibitors. These molecules mimic the nucleoside structure and become phosphorylated to the active drug, as explained in the previous example on acyclovir. The important difference here though is that the phosphorylation steps must all be carried out by cellular enzymes because, unlike herpesvirus, HIV does not have the necessary kinase enzyme.

The first drug to be approved for use against AIDS was zidovudine, a deoxythymidine mimic that was originally developed in the 1960s for treatment against cancer. The use of anti-nucleoside drugs as anti-viral medications is rational, with clearly understandable mechanisms of action in preventing DNA replication. The sugar $3'$ hydroxyl group is replaced by an azido group; it inhibits reverse transcriptase and also, on account of the azido group on the sugar ring, acts as a DNA chain terminator. Originally purposed as an anticancer agent, zidovudine was designed to stop the proliferation of cancerous cells by disenabling DNA replication needed for cell division. As might be expected for such a compound, zidovudine causes severe side effects, which ultimately led to its use being discontinued.

Indeed, the severe side effects of zidovudine, also known as AZT (azidothymidine), meant that the drug was very controversial during the late 1980s/early 1990s when it was approved for use against HIV/AIDS. At the time, people did not appreciate that AZT was a breakthrough in the treatment of HIV/AIDS and ultimately helped to destroy the notion that the disease was a death sentence. Since it was first realised that HIV was the cause of AIDS in 1984, concerted efforts to rationally design antiviral agents led to the utilisation of AZT, which was shown to be effective against HIV *in vitro* in 1985, and shortly after was approved for use against full AIDS in 1987, then approved for HIV in 1990. This was one of the fastest routes to clinic for any drug in the modern era. Campaigners argued that the early closure of the clinical trials was evidence that AZT was dangerously toxic and that this is why it failed cancer trials. This of course was not true; it is standard practice to give all patients a therapy once it is shown to save lives in a statistically significant way.

Besides the problems of associated side effects, viral resistance to AZT also became a problem, particularly if AZT was not used as part of a combination therapy. Consequently, it became a matter of urgency to develop new anti-HIV drugs, designed to avoid the problems of AZT resistance and improve on its performance. This led to a range of nucleoside-based reverse transcriptase inhibitors being introduced to the market. Understanding the structure–activity relationships with the target enzyme enables less toxic analogues to be developed, such as lamivudine, where the 3′ carbon is replaced with a sulphur atom which serves as a chain-terminating group. The structures of AZT and lamivudine are compared with the nucleoside thymidine in Figure 6.9.

Non-nucleoside reverse transcriptase inhibitors have a different mode of action. The molecules are generally hydrophobic and bind to a hydrophobic allosteric binding site, a binding site on the enzyme that is different to that used by the normal substrate, namely the active site. The physical effect of binding to a protein changes its shape, therefore deforms the active site, rendering it unusable by the normal substrate. NNRTIs are referred to as non-competitive reversible inhibitors because they are not in conflict with the normal substrate for access to the active site and their binding to the reverse transcriptase is not permanent. These two binding sites are distinct, so NNRTIs can be used in combination with NRTIs, which is advantageous because mutation can rapidly change the allosteric binding site and without a multi-drug approach would result in the preservation of resistant viruses.

The first-generation NNRTIs were discovered through random screening programmes. Nevirapine has a rigid butterfly-like conformation, making it chiral, where one 'wing' forms hydrophobic and Van der Waals interactions with aromatic residues in the binding site, while the other wing interacts with aliphatic residues. Another first-generation drug is delavirdine, which was developed from a lead compound that

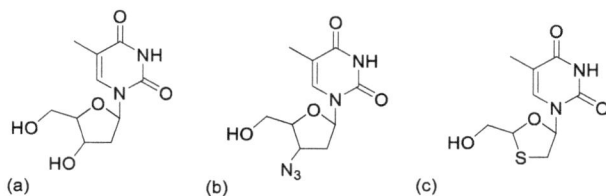

FIGURE 6.9 Molecular structures of the nucleoside thymidine (a) and nucleoside mimicking anti-HIV drugs AZT (b) and lamivudine (c).

emerged from a screening programme of 1500 structurally diverse compounds. Its structure differs from that of nevirapine most noticeably by the large tail that extends outside the normal binding site and projects out into the surrounding solvent. The alkyl pyridine ring motif is preserved, where hydrophobic contacts are important for binding to the same amino acids in the binding site. Unlike other first-generation NNRTIs, there is hydrogen bonding to residues in the main peptide chain of the binding site.

Changes to the amino acids in the binding site that form non-covalent interactions with the drug led to resistance against the first-generation NNRTIs. Crystallographic studies of the binding interactions and amino acid sequencing revealed that often a large amino acid is replaced by a smaller one, associated with the loss of an important binding interaction. Second-generation drugs were rationally designed to combat resistant viruses. Efavirenz is one such second-generation drug, developed as a result of comprehensive research on structure–binding relationships deduced from X-ray crystallographic interrogation of drug-binding site complexes of the first-generation drugs. X-ray crystallography and molecular modelling led to the structure-based design of a series of protease inhibitors. Multidrug approaches, using a combination of reverse transcriptase inhibitors together with PIs, are the most successful method of treating HIV infections because activity is improved, and preservation of resistant mutations is less likely.

Reverse transcriptase inhibitors are prodrugs and are activated by phosphorylation *in vivo*, but this is not the case with protease inhibitors. This is advantageous because it means that *in vitro* assays can be performed, where antiviral activity can be tested on infected cells in a petri dish, without the need for a patient or subject to be recipient of the untested drug, which has obvious ethical implications. Also, the protease enzyme can be isolated, and the PIs can be studied directly by X-ray crystallography to determine structure–activity relationships, therefore an effective drug can be readily designed.

In this way, important tests can be carried out without a risk to life and an effective drug can be designed, then used to carry out the necessary tests, such as IC_{50}, which gives an indication of how effective the protease inhibitor will be. The IC_{50} is the blood concentration of the drug required to inhibit 50% of the enzyme, so the lower this value is, the more potent the drug. However, a strong inhibitor does not necessarily correspond to good activity. In order to be effective, the drug must be able to pass the cell membrane of infected cells. Hence this is why *in vitro* whole cell assays are used alongside enzyme studies to ensure good absorption. Isolated lymphocytes infected with HIV are treated with the novel drug; antiviral activity is measured and recorded as an EC_{50} value, the concentration required to inhibit 50% of the cytopathic effect of the virus.

Good oral bioavailability is another important issue because anti-HIV drugs need to be taken for the duration of the patients' life, so a convenient route of administration is imperative. This is a particularly problematic aspect of PI development. PIs are developed from peptide lead compounds. Peptides notoriously have poor pharmacokinetic properties: poor absorption, metabolic susceptibility, rapid excretion, limited access to CNS and high plasma protein binding. This is due to high molecular weight, poor water solubility and the susceptibility of peptide bonds to hydrolysis.

The HIV protease enzyme is an aspartyl protease, which contains an aspartic acid residue in the active site, which is crucial for the catalytic cleavage on peptide bonds. The enzyme is a relatively small protein that can be readily made by synthetic techniques or by cloning and expression in rapidly dividing cells, then isolated and purified in large quantities. Crystallisation of HIV protease is relatively straightforward; hence this enzyme is an ideal target for rational structure-based drug design. From X-ray crystallographic studies, novel inhibitors can be developed to produce promising lead compounds.

The HIV protease enzyme is a dimer composed of two identical protein units, each consisting of 99 amino acids. The active site resides at the interface of the two units and, like the overall protein structure, is symmetrical about an axis of two-fold rotational (C2) symmetry. It has a broad substrate specificity, being able to cleave a variety of peptide bonds in viral polypeptides, but crucially it will hydrolyse bonds between proline and an aromatic residue (phenylalanine or tyrosine). This is an important feature because cleavage of these bonds is unusual and cannot be done by mammalian proteases. Furthermore, mammalian proteases lack the C2 symmetry characteristic of HIV protease. These features offer the possibility of drug selectivity.

HIV protease can be readily synthesised or isolated from cellular assays, then easily crystallised for study under X-ray diffraction and molecular modelling, as well as the differentiation from mammalian proteases, making this a good target for anti-HIV drug design. The design of HIV protease inhibitors was inspired by the extensive research done on the mammalian aspartic protease, renin. This enzyme was studied prior to the discovery of HIV protease and has a similar hydrolytic mechanism; hence its inhibitors have a close resemblance to those of HIV protease.

These agents act as transition state inhibitors, which mimic the transition state of the substrate during this enzyme catalysed reaction. This approach is advantageous because the transition state structure is more strongly bound to the active site than the substrate or products, so inhibitors that are structurally similar to the transition state will be most effective. As shown in the reaction mechanism in Figure 6.10, the transition state resembles the intermediate (2). However, the structure is inherently unstable, so an inhibitor must be designed that contains a transition state isostere, which mimics the tetrahedral centre of the transition state, but is stable to hydrolysis.

The importance of the dimer nature of the protease enzyme is evident in this mechanism as the two equivalent aspartate residues, Asp-25 and Asp-25′, work cooperatively

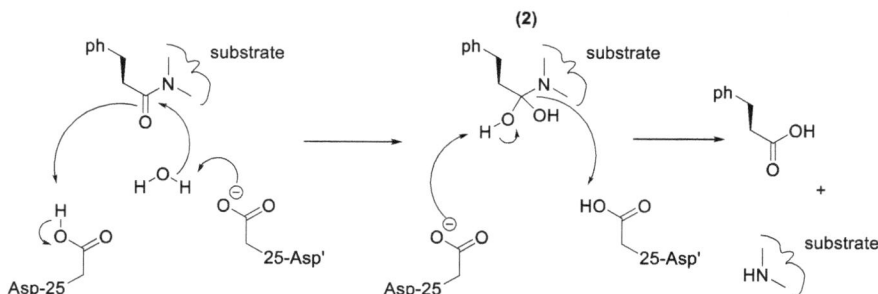

FIGURE 6.10   Mechanism of the enzymatic hydrolysis of the transition state isostere.

to catalyse the hydrolysis of the amide bond of the isostere shown explicitly here without convolution from drawing the rest of the substrate structure. This is an acid–base mechanism, using water as a nucleophile, which is activated by the Asp-25' residue. The TS intermediate (2) is an unstable structure, so the challenge for medicinal chemists was to incorporate this tetrahedral TS isostere into a molecular scaffold to produce a stable compound that would perform as an effective drug.

Several of these isosteres had been developed previously in the design of renin inhibitors. As a result, a huge library of compounds was synthesised incorporating these isosteres, with hydroxyethylamine proving to be the most successful. The effectiveness of this isostere can be attributed to the hydroxyl-substituted amine bearing close resemblance to the TS intermediate, and interacts with the aspartate residues in the active site in much the same way.

The next stage was to design protease inhibitors based on the enzymes' natural peptide substrate and to incorporate the isostere at the correct position. The first PI to be developed was saquinavir, by Roche. The natural peptide substrate was studied and the regions containing a phenylalanine-proline link were located, then a peptide sequence Leu[165]-Asn-Phe-Ile[169] was identified and served as the basis for inhibitor design. Since it is actually the phenylalanine-proline peptide bond that is hydrolysed, this link was replaced by the hydroxyethylamine TS isostere. Consequently, the usual hydrolysis reaction cannot occur and the substrate functions as a successful inhibitor. Additionally, the Leu-Asn-Phe-Ile residues are preserved in the substrate in order to bind effectively to protease subsites. Despite this, inhibition of the enzyme is relatively weak. Furthermore, high molecular weight and peptide-like characteristics are detrimental to oral bioavailability. As a result, Roche needed to find a smaller inhibitor based on the TS isostere that replaces the Phe-Pro dipeptide. An N- and C-protected structure of the hydroxyethylamine was tested first and was found to have weak inhibitory activity. The inclusion of an asparagine group to occupy one of the subsites resulted in a 40-fold increase in activity, greater than that of the original pentapeptide. This may seem paradoxical, as the latter occupies four binding subsites, but it has been found that the crucial interaction is that of the asparagine with its binding subsite. Other binding interactions to other subsites weaken this core interaction, therefore there is a decrease in activity.

From these findings, it was concluded that the hydroxyethylamine isostere bonded to asparagine had the greatest efficacy and was selected as the lead compound which led to saquinavir. The structure of saquinavir is shown in Figure 6.11. X-ray crystal analysis revealed that the protecting group occupies another subsite, which was shown to be a large hydrophobic pocket. The protecting group was replaced by a large quinolone ring system to optimise binding to this pocket, and this resulted in a six-fold increase in activity. Further modifications to the carboxylate end of the molecule yielded saquinavir, which had a further 60-fold activity compared to the prior lead compound.

Contemplation of the efficacy of the drug can be perceived from the properties of saquinavir. It is selective for HIV-1 and HIV-2 proteases, 100 times greater than for human proteases. Clinical trials were conducted in 1991 and the drug reached the market in 1995, indicative of the faith that medicinal chemists had in this medicine. However, about 45% of patients developed clinical resistance over a one-year period.

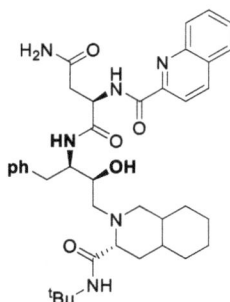

FIGURE 6.11   Molecular structure of saquinavir, an anti-HIV drug, with the TS isostere shown in bold.

Resistance is delayed if taken in combination with reverse transcriptase inhibitors, though. Oral bioavailability is poor, only about 4% in animal studies, but this can be improved when taken with meals. There is also 98% binding with plasma proteins, therefore the drug needs high dosing to achieve therapeutic plasma levels.

While saquinavir was an advancement from previous lead compounds, a more efficacious drug was required, with a lower molecular weight, less peptide character and better oral bioavailability. Merck had designed a potent PI by taking advantage of the symmetrical nature of the protease active site. By combining half of one PI with half of another, it would be possible to create a structurally distinct hybrid inhibitor. Chemists at Merck decided to combine the p′ half of their lead compound, which has less peptide character, with the p′ half of saquinavir because of its solubility enhancing potential. Lead optimisation led to the structure of indinavir, which reached the market in 1996. It had improved oral bioavailability (15%) and is less highly bound to plasma proteins (60%). This approach of combinatorial chemistry was extended to produce other ranges of PIs also.

Influenza is another example of an RNA virus. The rational design of drugs to treat this pathogen yielded different strategies to combat viruses to those used against HIV. Influenza causes the airborne disease commonly known as 'the flu' when it infects the epithelial cells of the upper respiratory tract. It is a major source of mortality, particularly among the elderly and persons with a weakened immune system. Most notable is the pandemic in 1918, where Spanish flu caused the death of at least 20 million people worldwide: greater than the number of casualties in the First World War. Further serious epidemics have since occurred, such as SARS, Bird flu and Swine fly, and it is likely that their origin is from China, where people live in close proximity to poultry and pigs, so the risk of cross-species infection is much higher.

Occasionally, and mistakenly, the terms 'flu' and 'cold' are used interchangeably, for example, the severe disease called 'man flu'. Note that the common cold is caused by a different kind of virus to influenza, called rhinoviruses, so different therapeutic agents are required to combat colds. Human rhinoviruses are among the smallest animal RNA viruses, containing a single strand of positive RNA enveloped by an icosahedral capsid composed of 60 protein units of which there are four distinct types. The structure of the influenza virus is very different.

The (−) single-stranded RNA of influenza is contained within the capsid, which itself is enveloped in an outer membrane constructed from that of the host cell and contains two viral glycoproteins: neuraminidase (NA) and haemagglutinin (HA), so-called because it can bind to red blood cells. The function of these two glycoproteins is to facilitate the infection process. NA helps the virus to traverse the mucus layers of the respiratory tract by catalysing degradation of the mucus layer, thus enabling the virus to reach the surface epithelial cells. Once there, adsorption of the virus can occur, whereby the virus binds to the host cell receptors that are recognised by HA, which binds to them rather than catalysing their degradation. The virion is now adsorbed; this process is known as receptor-mediated endocytosis. The pH then decreases inside the endosome, which is the membrane-bound compartment inside the host cell that contains the virion, causing HA to drastically change its conformation, where the hydrophobic ends of the protein fold outwards; extending towards the endosomal membrane and fusion occur on contact, enabling the RNA capsid to be released into the cytoplasm of the host cell. Disintegration of the capsid releases the RNA and the viral enzyme RNA polymerase, both of which invade the nucleus.

Viral RNA polymerase now begins to catalyse the copying of the viral (−) RNA as (+) viral RNA, which departs from the nucleus and acts as the mRNA needed for the translation of viral proteins. Capsid proteins made in the cell spontaneously self-assemble in the cytoplasm with incorporation of replicated strands of (−) RNA as well as newly synthesised RNA polymerase. The glycoproteins NA and HA are then incorporated before the virion is released from the cell by budding. Interactions of the glycoproteins NA and HA are crucial to this step. There needs to be a balance between the rate of desialylation by NA to aid the virion's departure from the cell and the rate of binding by HA, which was required for access to the cell. The amino acids present in the active site of NA are highly conserved, which illustrates the importance of the activity level of this enzyme.

The glycoproteins NA and HA are present on the outer surface of the virion, so can serve as antigens, potentially recognised by antibodies and the cells of the immune system. While the amino acids in the active site of NA and HA are highly conserved, the influenza virus is adept at varying the amino acid sequence elsewhere in the proteins: this propensity for mutation hinders recognition by the immune system and limits the ability to develop vaccines with these antigens. The reason for this high level of antigenic variation is because the RNA polymerase enzyme is highly error prone. This results in the production of RNA that codes for NA and HA being inconsistent. Variation in the RNA code means that different sequences of amino acids give different antigenic variants, resulting in different strains of flu. Where the variation is small, it is referred to as antigenic drift. When there is large variation, called antigenic shift, this can lead to more serious epidemics and pandemics.

As mentioned previously, developing a vaccine for influenza can present a challenge. For eventualities where vaccination proves unsuccessful, medicinal chemists must develop antiviral agents which remain effective despite antigenic drift. Given that neuraminidase is crucial to the infectious process, it presents a promising target for drug development.

Studies of the enzyme's crystal structure using X-ray diffraction and molecular modelling revealed that neuraminidase is a mushroom-shaped tetrameric glycoprotein, attached to the viral membrane by a single hydrophobic sequence consisting of

29 amino acids, which can be cleaved from the surface enzymatically to enable the unadulterated polypeptide to be studied without loss of antigenic or enzymatic activity. It was found that the active site is a deep pocket located centrally on each protein subunit, composed of 18 amino acids that are highly conserved. The outer structure of the protein is much more variable.

Since the amino acids in the active site are constant, and inhibition of the active site would severely affect the infectious process, neuraminidase inhibitors are an attractive target for medicinal chemists to create an antiviral therapy against influenza. By finding an inhibitor for the active site of NA, all strains of influenza would be targeted. Furthermore, the neuraminidase active site is unlike those of comparable enzymes for bacterial or mammalian cells, so there is no complication with selectivity.

During efforts directed towards elucidation of a structure for a possible inhibitor, NA has been crystallised with the product of the enzyme-catalysed reaction, sialic acid, bound to the active site in order to study the interactions. The most important interactions are hydrogen bonds and ionic interactions or the carboxylate group of sialic acid with three arginine residues, particularly with Arg-371. To achieve these interactions, sialic acid needs to be distorted from the most stable chair conformation, where the carboxyl group placed in an axial position on the chair changes to a pseudo-boat conformation, where the carboxylate ion is equatorial.

The nomenclature of these conformations is derived from vague observations of the different shapes that the cyclic molecules can adopt, described by a simile. The shapes are important because they predetermine the interactions that are possible between the molecule and active site. For instance, the carboxyl group placed in an axial position on a chair conformation of sialic acid has a different directionality to the equatorial position that is observed for the pseudo-boat conformation, so the hydrogen bonding requirements for each conformation are different and this has consequences for enzyme–substrate binding. Consequently, sialic acid adopts the pseudo-boat conformation because this maximises hydrogen bonding in the ES complex, despite this being the energetically least favourable conformation.

Other important binding regions include the interaction of a glycerol side chain of sialic acid with glutamate residues and a water molecule by hydrogen bonding. The hydroxyl group at C4 is situated in another binding pocket and interacts with a glutamate residue, and lastly, the acetamido substituent fits into a hydrophobic pocket, which is important for molecular recognition. The pocket contains the hydrophobic residues Trp-178 and Ile-222 which lie close to the methyl carbon C11 and the hydrocarbon backbone of the glycerol side chain. Furthermore, the distorted pyranose ring binds to the floor of the active site with its hydrophobic face and glycosidic OH at C2 is moved into an axial position where it can form hydrogen bonds to Asp-151.

By understanding the binding of the substrate, chemists were able to propose a mechanism for the hydrolysis reaction. It was discovered to involve proton donation from an activated water molecule, facilitated by the negatively charged Asp-151 and formation of an endocyclic sialosyl cation TS intermediate. Then sialic acid is formed and released. This proposed mechanism was supported by kinetic isotope studies, which indicate an S_N1 nucleophilic substitution, and NMR spectroscopy, a technique used to study the change in conformation of the substrate, showed that sialic acid is released as the α-anomer, consistent with an S_N1 mechanism. Also,

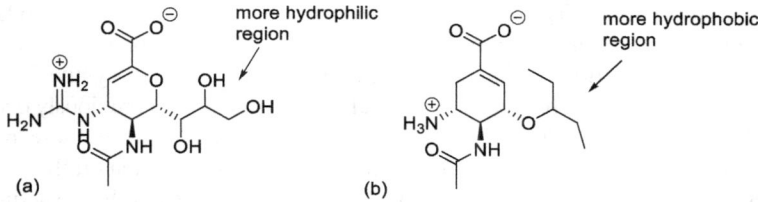

FIGURE 6.12   Molecular structure of anti-influenza drugs (a) Relenza™ and (b) Tamiflu™.

site-directed mutagenesis showed replacing the charged amino acids Arg-151 with lysine and Glu-227 by aspartate, so that stabilisation of the intermediate is repressed and activity of the enzyme is lost.

Once the structure of the TS and the mechanism of hydrolysis were thoroughly understood, companies could begin to develop inhibitors of the neuraminidase active site in an effort to generate antiviral agents against influenza. At the forefront of these drugs were Relenza and Tamiflu (Figure 6.12).

Sialic acid analogues were synthesised with a double bond between C2 and C3 to replicate the trigonal planar geometry of the transition state of the hydrolysis reaction, based on molecular modelling programmes to evaluate binding interactions in the active site. This approach used the molecular fragments techniques, where probe atoms are situated in the binding pockets within the active site and the strength of the binding interactions is measured via energy calculations using software. The probe atoms constitute important functional groups, such as the carboxylate group, ammonium cation and hydroxyl groups, as well as the hydrophobic methyl group. Hydrogen bonding exhibits directionality, so the orientation of these fragments needs to be comprehended in order to attain the most favourable interaction energy. The next stage is to develop a molecular scaffold that places the functional groups in such a way as to optimise these interactions.

Once these parameters were ascertained, the relevant structures were synthesised and tested for activity. Molecular modelling predicted that the active site of the enzyme that normally binds the 4-hydroxyl group of sialoside is different in the viral enzyme compared to comparable bacterial and mammalian enzymes, which was confirmed by the crystal structure of inhibitor-bound enzymes, so selectivity could be achieved by replacing the 4-hydroxy with a different functionality; the larger guanidinium group was found to have greater hydrogen bonding interactions. This led to the structure of Relenza, developed by GSK, which was found to be a potent inhibitor of influenza NA. However, there were limitations regarding oral bioavailability (<5%) attributable to the polar nature of the molecule. Consequently, this drug must be administered by inhalation.

Roche took a different approach to inhibition of neuraminidase. From studies of early sialic acid analogues, it was realised that the structure of the dihydropyran oxygen has no major role in binding to the active site, hence the possibility of replacing it with a methylene isostere to form a carbocyclic analogue; it would be advantageous to omit the polar oxygen atom, thus increasing the hydrophobicity of the molecule and potentially improve oral bioavailability. Furthermore, the glycerol side chain was removed to reduce polarity. This was replaced by a hydroxyl group to introduce an inductive electron-withdrawing effect on the carbocyclic double bond to alleviate its high electron density. Adding the hydroxyl group here facilitated the synthesis of ether groups on which hydrophobic substituents could bind to the pocket previously occupied by the glycerol side chain. As a result, a library of alkoxy analogues were screened and optimisation led to the development of Tamiflu, which could be taken orally.

Relenza and Tamiflu were developed separately, but have similar structure and non-covalent interactions, which might be expected, given that they were both rationally designed against the same target. The charged functional groups generate very strong ionic interactions with charged residues in the active site of NA, and polar functional groups produce strong non-covalent interactions. This strong binding of the drugs to the active site results in potent inhibition of NA. The formulation of Relenza was too hydrophilic to cross membranes in the intestine, so had to be delivered into the lungs directly, which can cause inflammation. Tamiflu formulation was more hydrophobic and orally active, with the ability to cross cell membranes more efficiently. However, Tamiflu needs to be taken within 24 hours of the flu starting to give the drug time to reach therapeutic levels in the lungs and give relief from symptoms.

The rational drug design of antiviral agents provides excellent examples of how medicinal chemists can strategically develop medicines against diseases by understanding the molecular basis of the disease and identifying appropriate targets. To combat viral infections, different strategies have been employed, with drugs developed against nucleic acid targets as well as protein targets. Rational drug design is not limited to producing therapies against pathogens: by understanding the biochemical processes that proceed within the human body, medicines can be made to treat derangements in the normal operation of the body, which lead to disease. Most notable perhaps is cancer research. By understanding the processes that lead to cancer, targets can be identified for drug design. Often, the targets for cancer therapy are on the DNA molecule. Indeed, many of the drugs targeting nucleic acids used in antiviral medicine originated from lead compounds for cancer therapy.

TUTORIAL QUESTIONS

1. Draw a mechanism for the intramolecular ring opening of the β-lactam ring, which is crucial to its mode of action in damaging bacterial cell wall and suggest modifications to the structure of penicillin that improve its function as an antibiotic.
2. Outline briefly how antibiotic resistance can arise and spread.
3. Outline one general strategy to limit the emergence of drug resistance, and propose two different general approaches that could be taken to further drug development should resistance to a drug emerge.
4. Lamivudine is used to treat HIV/AIDS. State its biological target, and briefly outline the mode of action of this drug.
5. Explain how the molecular structure of anti-influenza drugs Relenza™ and Tamiflu™ differ and how the differences impact upon the bioavailability of each drug.

7 Cancer and Chemotherapy

ABSTRACT

Cancer is a complex disease, meaning that there is no single cause for cancer. In order for cancer to occur in the body, several elements that control the regulation of the cell cycle have to have a malfunction. When this occurs, the cancerous cells can reproduce indefinitely and uncontrollably, leading to the development of a carcinoma. Due to the nature of cancer being a complex disease, with several causative factors, it is difficult to treat. Traditional chemotherapy is hazardous due to the low specificity towards cancer cells over normal cells. Such methods include the use of intercalating agents, alkylating agents, cross-linking drugs and drugs that target the oestrogen receptor. More recently, the frontiers of cancer research have examined the use of antibody-directed therapies, which aim to deliver the harmful, traditional therapies to the target area of the tumour, thus reducing the harmful effects to healthy cells.

Cancer is a leading cause of death worldwide, accounting for nearly 10 million deaths in 2020, or nearly one in six deaths (WHO). It has become one of the most prevalent diseases in modern times, largely due to ageing populations. One in four people are likely to get cancer at some stage of their life.

7.1 CANCER, A COMPLEX DISEASE

Cancerous tissue develops from normal cells that have lost the normal regulatory mechanisms that control cell growth and multiplication. The abnormal cells multiply rapidly and uncontrollably to form cancerous tissue: neoplasm or tumours, which often lose their specialised characteristics that differentiate them from other types of cells. If the cancer is localised, it is said to be benign. If the cancer cells invade other parts of the body and produce secondary tumours, a process known as metastasis, the cancer is defined as malignant. This is the life-threatening form.

Cancer is termed a complex disease; there is no obvious single cause and there are more than 200 different types of cancer, which result from different cellular defects, therefore it is difficult to develop treatments to cure cancer. Therapies for one type of cancer may not be effective at treating another form of the disease.

Cancer can be caused by environmental influences, such as carcinogenic chemicals, for example, in cigarette smoke or in certain foods, which induce gene mutations or interfere with normal cell differentiation and division. Genetic mutation in genes that are responsible for controlling cell division can result in uncontrolled multiplication of cells to produce a tumour. In this way, mutagenic chemicals can result in carcinogenesis, the development of cancer.

DOI: 10.1201/9781003516835-7

The involvement of viruses in at least six human cancers has been observed and is responsible for a significant proportion of the world's cancer deaths, though their involvement may not always be clear. Epstein–Barr virus causes Burkitt's lymphoma and nasopharyngeal carcinoma; human papillomaviruses are STIs, which can lead to cancer in the cervix; and hepatitis B may be involved in many cancers of the liver. Viruses can cause cancer in several ways. They may bring oncogenes into the host cell and insert them into the genome, or some viruses may carry one or more promoters of gene mutation.

An individual may have a genetic predisposition for a certain cancer. Defective genes can be inherited, increasing the risk of cancer in subsequent generations. There are numerous possible genetic faults that can lead to cancer. Proto-oncogenes are genes that normally code for proteins that are involved in the control of cell division and differentiation. If they become mutated into an oncogene, this may disrupt the normal function of the gene and without control mechanisms in place, the cell could become cancerous. For example, the Ras protein is involved in the signalling pathway leading to cell division, during the processes of mitosis and meiosis. In normal cells, this protein has the self-regulating ability to switch itself off. In the mutated gene, Ras loses this ability, and is continually active leading to uncontrolled cell division. This mutation is found to be present in around one-fifth of human cancers.

In the event that DNA becomes damaged in a normal cell, there are recognition mechanisms in place so that the damage can be detected, and halting DNA replication gives the cell time to repair the damaged DNA before the next cell division. If repair is not possible, the cell commits suicide, a process called **apoptosis**. Tumour suppression genes code for proteins that are involved in the processes of checking, repair and apoptosis. If one of these genes, such as TP53, which codes for a protein with the same nomenclature, P53, becomes damaged, the P53 protein, which is crucial for the repair mechanisms, no longer functions and defects in the DNA persist, being carried to the next generation of cells. As the extent of the damage in the tissue increases, it becomes more likely that the tissue will develop into a carcinoma.

Genetic defects can result in numerous possible cell abnormalities, all of which are associated with cancer. These include: abnormal signalling pathways, such as insensitivity to growth-inhibitory signals and abnormalities in cell cycle regulation, evasion of programmed cell death (apoptosis) and limitless cell division, known as immortality. Cancerous tissue has the ability to develop new blood vessels, a process known as angiogenesis, which is required to nourish the fast-growing tissue. Tissue invasion and metastasis can occur, where cancerous cells spread to other parts of the body and can propagate cancerous tissue in other organs.

Most, if not all of these abnormalities must manifest before a defective cell can develop into a terminal malignant growth, because a series of safeguards are in place to prevent adverse effects that may arise from a single cell defect. This explains why cancers can take many years to develop after prolonged exposure to harmful mutagens, such as asbestos or coal dust. Short-term exposure may cause a limited number of mutations, but cellular safeguards keep control over abnormal cellular activity. A lifetime of exposure to damaging mutagens, as with long-term tobacco smokers, who are frequently exposed to carcinogenic chemicals, overwhelms the safeguard mechanisms until the abnormal cell is liberated from the control mechanisms and becomes cancerous.

Cancers usually occur later in life as a result of exposure to mutagens over a longer timescale, and also because of age-related breakdown of control mechanisms. This is also one reason why cancer is difficult to treat: by the time cancer has appeared, several cellular defects have arisen and over-ridden multiple control mechanisms, so addressing one cell defect with a therapy is unlikely to be effective. Consequently, traditional anticancer agents tend to be highly toxic, acting on a variety of targets via different mechanisms: focussed on death of abnormal cells, but without high specificity over normal tissues. These drugs are referred to as being cytotoxic, so a dose rate must be carefully selected to ensure suppression of the carcinoma, but at the same time is bearable for the patient.

More recently, anticancer drugs have been developed that target specific abnormalities in cancer cells, improving selectivity and thus reducing side effects. However, cancer cells will contain multiple cell defects. Targeting one will not be an effective treatment; hence combination therapy is required, with drugs that target cancer via different mechanisms. This approach may be used in conjugation with radiotherapy or surgery.

7.1.1 Cancer and the Cell Cycle

Abnormal signalling pathways disrupt normal cell growth and division. The normal signalling pathway involves hormones, known as growth factors, which are extracellular chemical messengers which activate protein kinase receptors in the cell membrane. These receptors trigger a signal transduction pathway which eventually reaches the nucleus and promotes transcription of proteins and enzymes needed for cell growth and division. Defects in this signalling process result in the cell being constantly instructed to multiply, ultimately resulting in cancerous growths. The complexity of this signalling process means that there are several points of the cell cycle that can go wrong.

Furthermore, many cancer cells are capable of growing and dividing in the absence of external growth factors because they have the ability to produce their own growth factors; releasing them stimulates the cell's own receptors. Examples of growth factors include platelet-derived growth factor (PDGF) and transforming growth factor-α (TGF-α), which are important in the development of a carcinoma. Cancer cells may also produce abnormal receptors, which are constantly switched on, despite the absence of growth factors. Also, it is possible for a receptor to be overexpressed when an oncogene is too active, coding for excessive quantities of protein receptor in the cell membrane, which means that the cell becomes supersensitive to low levels of growth factor. For example, Ras protein receptors in cancerous cells can be over-abundant and usually lose the ability to autoregulate, being constantly switched on and the signalling pathway for cell division becomes over-active.

Additionally, there are several external hormones in place, such as transforming growth factor-β (TGF-β), that counteract the effects of stimulatory growth factors and signal for the inhibition of cell growth and division. Damage to the genes that code for the receptors of these inhibitory hormones, the tumour suppressor genes, can result in insensitivity to these signals and increase the risk of cells becoming cancerous.

Chromosomes containing mutated genes will become more prevalent if cell division and replication of the defective DNA are allowed to continue unimpeded, when the necessary control mechanisms fail, and this vastly increases the risk of cancer; as defective cells multiply in number, they can grow uncontrollably and defective genes can be inherited in gametes, if damage occurs during meiosis. Humans have 22 pairs of homologous chromosomes (X and Y) and a pair of sex chromosomes: a total of 46 chromosomes, known as the diploid number. The diploid number varies between species, for example, dogs have 78. These are always even numbers because half comes from each parent. The chromosomes are dispersed throughout the nucleus, and each comprises a single molecule of DNA that is supercoiled, held in place by protein molecules called histones. The chromosomes are only distinctly visible under the microscope during cell division.

Before cell division can occur, DNA replication must take place so that the genetic material can be passed to the daughter cells. DNA unwinds ready for replication, then the two strands are 'unzipped' by the enzyme DNA helicase, which breaks the hydrogen bonds between bases. The exposed bases of each strand can now undergo complementary base pairing with free nucleotides in the nucleus, which are assembled by another enzyme, called DNA polymerase. Finally, all the nucleotides are combined to form two new polynucleotides, each comprises an original strand and a new strand. This is called semi-conservative replication.

The daughter cells, produced after cell division, will now have the correct genetic information for controlling cellular processes. There are two forms of cell division: mitosis and meiosis. **Mitosis** produces two daughter cells and enables cells to replicate for growth and repair. **Meiosis** produces four daughter cells, each with half the number of chromosomes: haploid cells, which are important for making gametes, which fuse during fertilisation to make diploid cells, and ultimately a foetus.

Mitosis proceeds in five phases. During interphase, the cell is in its normal form; not dividing and the nucleus contains loose DNA, chromatin. During prophase, the nuclear envelope disappears, and chromosomes become visible. The chromosomes arrange themselves at the equator (centre) of the cell in metaphase, attached to spindle fibres. In anaphase, each of the two threads of a chromosome (chromatid) migrates to the opposite pole. By telophase, the nuclear envelope reforms and the cell splits to yield two identical diploid cells. If mutation occurs, the daughter cells will not be genetically identical to the parent cell and risk not functioning properly in the body.

Cells undergo a regular cycle of cell division, which is split into three periods. Interphase comprises the majority of the cell cycle, when there is no division, and is subdivided into a further three parts: first growth phase (G1), where organelles produce proteins; synthesis phase, when DNA is replicated; and second growth phase (G2), when organelles grow and divide, and energy stores are increased. The second phase is nuclear division, where the nucleus divides into either two (mitosis) or four (meiosis). The last phase is cell division to produce the daughter cells. For mammals, the cell cycle takes about 24 hours, 90% of which is interphase. Most cancers are caused by damage to the genes that regulate mitosis and the cell cycle.

Abnormalities in cell cycle regulation may occur at any of the four major phases, known as G_1, S, G_2 and M. Progression through these phases of the cell cycle depends on the balance of the chemical signals that promote growth or inhibition. The G_1

(gap 1) phase is where the cell grows in size and prepares for DNA replication in response to growth factors. The second phase (synthesis) is when DNA replication takes place. The next interval, once the chromosomes have been copied, is G_2 phase (gap 2) where the cell prepares for division. During this interval, the cell has time to check for errors in the DNA replication and repair any damaged copies. The final phase, M (mitosis), is when cell division happens to produce two daughter cells, each containing a full set of chromosomes. The daughter cell then begins the cell cycle at the G_1 phase, or may remain in a dormant phase G_0.

During the cell cycle, there are various decision points to determine whether the cell should continue to the next phase or not. The restriction point, R, during the G_1 phase frequently becomes abnormal in cancer cells. Also, surveillance mechanisms are in place, known as check points, which assess the integrity of the process and a delay will occur in the G_2 phase if damage to DNA is detected. During this time, repair mechanisms can work to rectify the damage, or if the damage incurred is too great, the cell may commit suicide during apoptosis. If these checkpoints fail to operate properly, cancer can arise. The replication of malfunctioning cells, which lack the necessary mechanisms of quality control, can lead to the development of a tumour because there is no mechanism in place to stop their uncontrolled division.

Among the mechanisms in place to control the cell cycle are cascades involving a variety of proteins called cyclins, along with enzymes called cyclin-dependent kinases (CDK). Different types of each are responsible for the regulation of the different phases of the cell cycle. Binding of a cyclin with its enzyme activates the enzyme and serves to move the cell cycle from one phase to the next. Progression through the cell cycle is regulated by sequential activation of cyclins and CDKs. The process can also be downregulated by CDK inhibitors.

A number of restraining proteins are in place to regulate the effects of cyclins and have an inhibitory effect by blocking the activity of CDKs. The amount of restraining protein present is controlled by P53, which is important for monitoring the health of the cell and the integrity of DNA. Several cancers are associated with overactive cyclins or CDKs. This occurs when these molecules are produced in excess, or the genes that code for inhibitory proteins are malfunctioning. Indeed, half of all human tumours lack a properly functioning P53 protein, consequently the levels of restraining protein are low; in other words, there is insufficient production of CDK inhibitors and so the cell cycle proceeds unregulated.

The P53 protein also has an important role in apoptosis. Under normal circumstances, faulty cells that cannot be repaired have an intrinsic self-destruction mechanism, which is automatically initiated if the monitoring processes at checkpoints fail to recognise the necessary chemical signals that divulge corrupt DNA.

There are two distinct pathways for apoptosis. One: an extrinsic route, where three external factors govern whether or not apoptosis occurs: these are sustained lack of growth factors/hormones, T lymphocytes of the immune system circulate the body and remove damaged cells, and death activator proteins, which may bind to tumour necrosis factor receptors on the cell membrane, triggering a signalling process that results in apoptosis. Two: an intrinsic pathway, where mechanisms within the cell detect damaged DNA and cause the production of more P53 tumour suppressor protein, which at sufficient levels triggers apoptosis.

Cancer cells can be described as 'immortal' because there is no apparent limit to the number of times that they can divide. Normal cells have a predetermined lifetime, based on the number of times that their DNA can be copied, which is typically 50–60 cell divisions. Structures called telomeres have a key role in the immortalisation process. The purpose of a telomere is to serve as a splice at the end of a chromosome and to stabilise and protect the DNA. It consists of a polynucleotide region at the 3′ end of the chromosome consisting of several thousand repeats of a short (six base pairs) sequence. Around 50–100 base pairs are lost from the telomere after each replication because DNA polymerase is unable to replicate the 3′ ends of the chromosomal DNA. Ultimately, the telomere becomes too short to function and the DNA becomes unstable and either unravels or links up end-to-end with another DNA molecule. This is a fatal error for the cell and apoptosis is triggered.

The immortality of cancer cells, the ability to divide an indefinite number of times, is due to the ability of cancer cells to maintain the length of their telomere by expressing an enzyme called telomerase. This is a type of RNA-dependent DNA polymerase, which works by adding hexanucleotide repeats on to the end of telomeric DNA, thereby maintaining its length. The gene encoding this enzyme is essential for the development of an embryo but is suppressed after birth. Cancer cells develop to remove this suppression and express the enzyme, achieving immortality as a consequence. It is found that telomerase is expressed in >85% of cancers.

Angiogenesis is the process by which new blood vessels are formed. As tumours grow, they require more essential nutrients, such as amino acids, nucleotides, carbohydrates as well as oxygen for respiration, so the tumour must have a good blood supply. The larger a tumour becomes, its cells, particularly towards the centre of the mass, become more remote from the blood supply, and so are starved of these resources. Tumour cells release growth factors to counteract this problem. Vascular endothelial growth factor (VEGF) and fibroblast growth factor (FGF-2) bind to cell receptors of the endothelial cells of nearby blood vessels and simulate the cells to divide, resulting in branching and extension of the existing capillary network to supply the ever-increasing demands of the tumour.

In normal cells, vascular growth factors are usually released when tissues have been damaged. Angiogenesis helps repair the injuries and is regulated by angiogenesis inhibitors, such as angiostatin and thrombospondin. Matrix metalloproteinases are enzymes that break down the membrane around the blood vessels to allow endothelial cells to migrate towards the tumour and enable the release of angiogenesis factors. Tumours are able to receive the blood supply they require due to a disruption to this balance. Furthermore, this increases the risk of cancer cells escaping from the primary source and metastasising because of the increased availability of blood, and newly developing endothelial cells can release proteins such as interleukin-6 that stimulate metastasis. The angiogenesis promoted by cancer cells results in the production of abnormal blood vessels, which are dilated, permeable and have a disorganised structure. Also, integrin molecules are displayed on the cell surface membrane, which are usually absent from mature cells, thus preventing apoptosis.

One possible way to combat cancer would be to inhibit angiogenesis. Drugs developed to break down abnormal blood vessels are likely to be less toxic than traditional chemotherapeutic agents but need to be used along with other treatments. The angiogenesis inhibitors serve to 'normalise' the tumour capillaries before destroying them. This helps to gain access to the tumour with other anticancer agents, and stall cancer growth and shrink the tumour because fewer capillaries are present to supply the tumour with the essential nutrients to grow.

The selective delivery of anticancer agents can be achieved by taking advantage of the leaky capillaries produced by angiogenesis. Anticancer drugs encapsulated into liposomes or nanospheres cannot escape normal blood vessels but can diffuse through the faulty ones in a tumour, so the anticancer drug accumulates in the tumour. Cancerous tissues tend not to develop an effective lymphatic system, so the polymeric drug delivery system becomes trapped at the tumour site, therefore when the drug is released, it is concentrated in the tumour.

In well-developed tumours, capillaries do not access all regions of the tissue, despite unregulated angiogenesis. These cells are deprived of oxygen and nutrients and may become dormant. This poses an issue for cancer therapy because most anticancer drugs work best on actively dividing cells. Even if a treatment has successfully eliminated the cancer, dormant cells start to multiply, and the cancer reappears. Moreover, these cells are more likely to metastasise.

Tissue invasion and metastasis can result in cancer spreading to other parts of the body; benign tumours are localised growths that remain in a particular part of the body, so are not life-threatening. By contrast, malignant cancer has the propensity for its cells to break away from the primary tumour and invade a blood vessel or the lymphatic system, be circulated and propagate a tumour elsewhere in the body, ultimately with fatal results. To do this, cancer cells must overcome the constraints that are designed to keep cells in place.

To ensure that cells remain in the correct part of the body, they have a molecular signature on their surface with which they adhere to other cells of similar character and to the insoluble extracellular protein matrix, which holds the tissue together. Cells of the tissue are anchored to the extracellular matrix with the aid of integrin proteins. In the event that a normal cell becomes detached, apoptosis is triggered, and this prevents cells from straying from one part of the body to another. Moreover, the adhesion molecules on the surface, the integrin proteins, must match the relevant extracellular matrix in order for the cell to adhere and survive.

When cancer cells metastasise, the integrin molecules are absent. Inevitably, they detach from the primary tumour, yet are able to survive independently and can attach to the extracellular matrix in another part of the body. Apoptosis is not triggered, and the immigrant cells can propagate a secondary tumour. It is hypothesised that the oncogenes code for false protein messages that imply that the cell is still attached.

Cancer cells and white blood cells are the only types of cells that are able to breach the extracellular matrix; both possess the metalloproteinase enzyme that hydrolyses the proteins that comprise this barrier. Once this barrier has been breached, cancer cells are free to penetrate blood vessels and enter the blood supply. The cancer cells then circulate in the blood and usually become trapped in the first network of capillaries that they encounter; for most tissues secondary tumours establish in the lungs, intestines or liver.

7.2   INTRODUCTION TO CANCER CHEMOTHERAPY

Among the doctor's arsenal for the treatment of cancer is surgery, radiotherapy and chemotherapy. Often a combination of these therapies is implemented, but for the medicinal chemist, chemotherapy is the pinnacle of cancer research. A variety of anticancer drugs may be used in combination, having different mechanisms of action in order to improve the likelihood of a successful treatment. Traditional anticancer treatments often affect normal cells as well as cancer cells due to their similarity, which makes it difficult to identify discrete targets, hence the renowned toxicity of chemotherapy. The challenge for the medicinal chemist is to develop a drug that will work on a specific target, or at least to devise a method of drug delivery that ensures the drug becomes concentrated in the tumour, thus minimising side effects.

Due to the nature of cancer cells rapidly dividing, fast growth, the large requirement of nutrients and the large network of capillaries to supply them, drugs do naturally tend to accumulate in cancerous tissue. However, a disadvantage is that dormant cells can evade the drugs, while normal cells that divide rapidly, such as bone marrow cells, can accede to the effects of the drug. This can lead to a weakening of the immune system because the chemotherapy impacts the production of white blood cells. Patients therefore have less resistance to infection, making the individual susceptible to pathogens and prone to infections.

Most traditional anticancer agents are cytotoxic; they work by disrupting the function of DNA in order to destroy abnormal cells. Some act directly on DNA while others, known as antimetabolites, affect the enzymes involved in DNA replication. Cancer research has led to a better understanding of the molecular mechanisms of abnormal cell chemistry. Knowledge of the cell cycle and genetic analysis have aided the development of new treatments, such as less toxic forms of chemotherapy, the use of monoclonal antibodies and enzyme prodrug therapy. Progression in anticancer research is now directed towards anticancer agents that act on specific molecular targets that are presented by cancer cells.

A challenge for cancer treatment is the occurrence of resistance, which can be intrinsic or acquired. Intrinsic resistance refers to a scenario where the anticancer drug is not effective at any stage of the treatment. This could be due to biochemical or genetic reasons, slow growth of the tissue, or poor uptake of the drug, for example, by dormant cells. Acquired resistance is where the drug initially is effective against the tumour, but the cancer cells develop to lose their susceptibility to the drug. This often results from the destruction of drug-sensitive cells to enrich the tumour in resistant cells present in the tumour, which continue to replicate and multiply. Ultimately a tumour grows that does not respond to the chemotherapy. This property of the tumour derives from genetic mutation of some cancer cells attributable to their genetically unstable disposition.

Resistance can arise through several molecular mechanisms that stem from mutation. Notably, the predominant avenues for tumour resistance include: decreased uptake of the drug, for example, by dormant cells; increased synthesis of a given target molecule beyond the scope of the therapeutic index of the drug; changes in metabolic pathways, meaning that the drug is no longer metabolised and therefore may no longer activate the prodrug. Furthermore, efflux pumps, p-glycoprotein transport channels, can expel a range of molecules and can result in multidrug resistance, hence efforts are underway to make p-glycoprotein inhibitors an active area in cancer research.

Different strategies can be employed to target DNA, producing drugs that work by different mechanisms to help regress the advancement of cancer. Intercalating agents and alkylating agents target DNA directly by distorting the shape of the helix and cross-linking the two nucleic acid strands, respectively. This has the result of interrupting DNA replication and suppressing cell multiplication. Alternative strategies involve targeting the enzymes involved in DNA synthesis; hormone therapy antagonises the receptors that promote DNA replication, and structural proteins that are important for mitosis can also be targeted.

7.2.1 Intercalating Agents

Certain natural products extracted from microbes were found to have an impact on the rate of cancer growth. These compounds acted as intercalating agents. The molecule inserts into the grooves along the sugar-phosphate backbone and binds by non-covalent means, causing a distortion to the shape of the helix. This has the effect of preventing the enzymes involved in DNA replication and transcription from operating on the molecule. Intercalators are flat aromatic compounds, which can fit into the grooves of the DNA helix, where heteroatoms form hydrogen bonds between the base pairs.

An example of a naturally occurring **intercalating agent** is doxorubicin; isolated from *Streptomyces peucetius*, it belongs to a group of antibiotics called anthracyclines. It has a tetracyclic ring system, where three of the rings are planar and intercalate the DNA double helix in the major groove. The charged amino group on the sugar ring forms an important ionic bond with the negatively charged phosphate groups of the DNA backbone, crucial for the drugs activity.

The mechanism of activity against DNA is intercalation-mediated topoisomerase inhibition, which prevents the normal function of the topoisomerase II enzyme. As the DNA double helix unwinds during the replication process, increased tension and entanglement can occur further down the helix. Where two sections of DNA are in close contact, the enzyme will bind to one of the double helices and a tyrosine residue in the active site is used to score both strands of the DNA by attacking a phosphate and cleaving the DNA backbone. This produces a temporary covalent bond between the 5′ end of the stand and the nucleophilic tyrosine oxygen of the enzyme, thereby stabilising the DNA. The intact section of DNA can now pass over the other strand, avoiding entanglement. The enzyme then reconnects the strands and departs.

With the intercalation of doxorubicin, the DNA–enzyme complex is stabilised such that the covalent bond to tyrosine is not cleaved and the enzyme cannot depart. This can lead to numerous DNA breaks, which, when encountered by DNA polymerases and helicases during the replication process, may trigger apoptosis. This is effective against rapidly proliferating cancer cells, where topoisomerase II is very active.

Mitoxantrone is a synthetic analogue to the anthracyclines, with a similar mechanism of action, used in the treatment of leukaemia and lymphomas predominantly. Its structure is simplified to allow for an uncomplicated synthesis, and lacks the sugar moiety of the anthracyclines, which is thought to cause cardiotoxic side effects. However, the amino group of the sugar moiety is important for binding, and is needed to be replaced by a suitable substituent that would place the required nitrogen

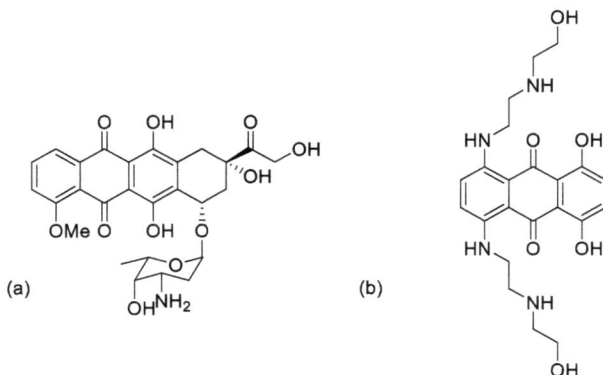

FIGURE 7.1   The anthracycline drug doxorubicin (a) and synthetic analogue mitoxantrone (b).

in the same relative position. The tetracyclic ring system was truncated by removing the non-planar ring and replacing it with two identical substituents to make the molecule symmetrical and therefore easier to synthesise. The planar tricyclic ring system intercalates the DNA in much the same way as the anthracyclines, with studies of structure–activity relationships revealing the carbonyl and phenol groups to be an important pharmacophore, while the amine nitrogen is crucial for binding. The structures of the anthracycline drug doxorubicin and synthetic analogue mitoxantrone are given in Figure 7.1.

7.2.2 ALKYLATING AGENTS

Mustard gases were used in the First World War to devastating effect. It was later realised, however, that nitrogen mustards had potential as anticancer agents. They are classified as **alkylating agents**, referring to their mechanism of action. These compounds are highly electrophilic and readily react with nucleophiles to form strong covalent bonds. Alkylating molecules will react with the nucleophilic nitrogen groups of DNA bases, particularly N-7 of guanine. Symmetrical molecules are used, with electrophilic groups on either side, which will react with guanine bases on each strand of DNA and therefore cross-link them, which causes transcription and translation to be disrupted and therefore impede the progress of cancer. Conversely, the drug may link to two guanine bases on the same strand of DNA and cover that portion of the DNA from the actions of enzymes involved in replication. Predictably, there are limitations to alkylating drugs. Such reactive compounds will attack other nucleophilic groups on proteins also, so have poor selectivity, and themselves are mutagenic due to their action on DNA.

The first alkylating agent to be used medicinally was chlormethine, in 1942. Its mechanism of action, as shown in Figure 7.2, involves displacement of a chloride ion via intramolecular nucleophilic substitution to produce a highly electrophilic aziridinium ion. Alkylation of DNA can then occur when the nucleophilic N-7 of guanine attacks the β-carbon of the electrophile. This process is repeated with the other strand of DNA to cause cross-linking, halt DNA replication and thus serve as an anticancer agent.

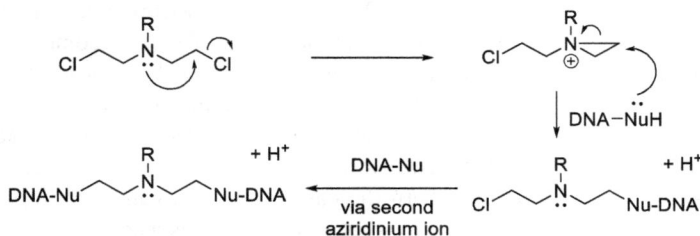

FIGURE 7.2 Mechanism for the cross-coupling of DNA by nitrogen mustards.

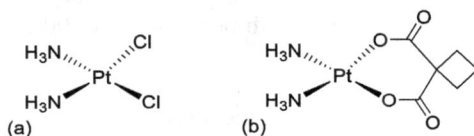

FIGURE 7.3 Structure of the platinum complex known as cisplatin (a) and the second-generation analogue carboplatin (b).

The nitrogen mustards are highly reactive and will react with other nucleophiles besides guanine. They are far too reactive to survive the oral route, so must be administered intravenously. In order to modulate activity and reduce unwanted side reactions, electron-withdrawing substituents can be added to the nitrogen atoms in place of Me, such as an aromatic ring. This has the effect of making the lone pair of electrons on the nitrogen less available for displacing the chloride ion and the rate of formation of the aziridinium ion is reduced. In this example, the lone pair interacts with the resonating electrons of the aromatic ring, so only strong nucleophiles like N-7 of guanine are able to force the alkylation reaction.

Strategies can be adopted to improve the selectivity of alkylating agents. Attaching a moiety to the aromatic ring to mimic the structure of phenyl alanine, so that the drug is recognised as an amino acid, allows the drug to be readily taken into cancer cells, where the demand for amino acids is greater, via transport proteins. A similar approach is used with uracil mustards, where the nucleic acid building block uracil is attached to the mustard. Again, selectivity is displayed as a consequence of the greater demand for nucleic acid bases by tumours compared to normal tissue, so higher concentrations of the drug accumulate in the tumour.

7.2.3 CROSS-LINKING DRUGS

Metal complexes have been used as **cross-linking drugs** to target DNA directly. Cisplatin is frequently used in the IV treatment of testicular and ovarian cancer. Its discovery was fortuitous from research in the 1960s on effects of electric currents on bacterial growth. It was found that bacterial growth was inhibited by an electrolysis product from the platinum electrode: cisplatin. The structure of cisplatin is shown in Figure 7.3.

Cisplatin acts primarily as a DNA 1, 2-intrastrand cross-linker, and its molecular geometry is key to its chemical mode of action. The molecular geometry, which is flat, referred to as square planar, arises from the nature of the bonding that is present between metals and their ligands. For the organic molecules considered so far in this text, covalent bonding involved shared pairs of electrons spaced at the furthest distance possible to yield a tetrahedral geometry, with bond angles of $109.5°$, as predicted by the VSEPR model. In the case of metal–ligand bonding, both the electrons are supplied by the ligand in what is referred to as a coordinate bond. This occurs because metals have a different valence electron configuration to other types of atoms, with defined energy levels that allow for this type of bonding to predominate. In the case of cisplatin, the ammonia and chloride ligands each donate their lone pair of electrons to the platinum ion, and the occupancy of these electron shells dictates the square planar geometry, with bond angles of $90°$. The geometry of the metal complex is key to the mode of action.

The chloride ligands are displaced by a water molecule *in vivo*, and the $[Pt(NH_3)_2(H_2O)]^+$ complex approaches the negatively charged DNA. The water ligand is replaced when Pt forms a coordinate bond to N-7 of the nucleobase guanine, causing distortion of the purines and the DNA helix becomes kinked. As a result, DNA replication and repair cannot occur, and the cell dies. Cisplatin undergoes this process very aggressively, and consequently leads to severe side effects. Attempts to produce analogues that are less toxic involved the utilisation of organic ligands. The organic molecules in place of the two chlorides have two donor groups, so two coordinate bonds are formed with one ligand, making the ligands more difficult to displace by a water molecule, hence the activity is reduced.

7.2.4 TARGETING THE OESTROGEN RECEPTOR

Another approach to impede DNA replication is to target enzymes involved in the synthesis of DNA. By interfering with the actual biosynthesis of DNA, the growth of tumours can be reduced, but this approach must be targeted and monitored carefully to avoid undue damage to normal processes. Folic acid (vitamin B9) plays an essential role in the biosynthesis of nucleic acid bases, so presents a prudent target for finding antagonists.

Some cancers are hormone-dependent. Hormones can serve as a signal for DNA replication, and derangements in this signalling process may lead to cancer. Steroid hormones bind to intracellular receptors to form complexes that act as nuclear transcription factors to control whether transcription takes place. Hormone antagonists can be used to block these receptors, thus impede cell proliferation. Oestrogen is commonly involved in these pathways. Oestrogen drugs are used in the treatment of prostate cancer because they inhibit the production of luteinising hormone and thereby decrease the synthesis of testosterone, which is linked with prostate cancer. In hormone-dependent breast cancer, oestrogen antagonists are used to block oestrogen receptors and prevent hormone binding.

Figure 7.4 demonstrates how closely the structure of stilboestrol resembles that of oestrogen if bonds are imagined between the ethyl groups and the 3-position of the phenol rings, therefore it is unsurprising that stilboestrol acts as an agonist in the

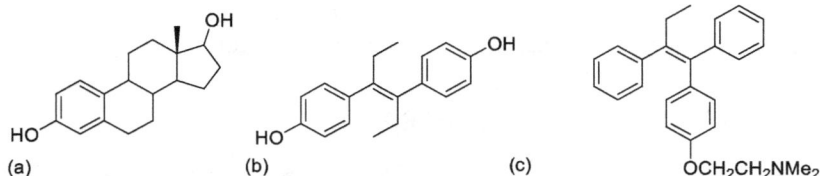

FIGURE 7.4 Molecular structures of oestrogen (a), stilboestrol (b), and tamoxifen (c).

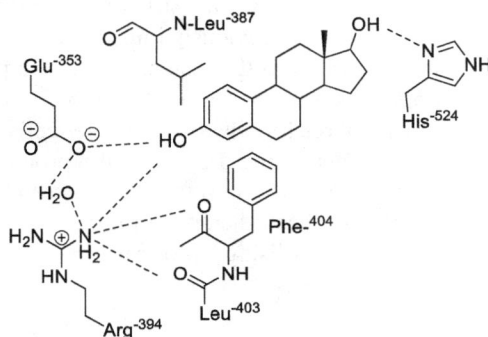

FIGURE 7.5 The binding of oestrogen with key residues in the oestrogen receptor.

oestrogen receptor. The structure of tamoxifen is also similar to the extent that it can bind to the oestrogen receptor but is sufficiently different not to initiate a response upon binding, therefore acting as an antagonist. Note that the E-stereochemistry of the alkene is important in stilboestrol for binding, and indeed the E-isomer of tamoxifen acts as an oestrogen agonist also. The OH group is also important for binding. The two hydroxyl groups are separated by what is essentially a hydrophobic spacer so that they are in the correct position relative to each other to fit specifically into the oestrogen receptor site. The two hydroxyl groups nestle into polar pockets within the receptor: crucially the phenolic group forms hydrogen bonds with three groups. Molecular shape complementarity arises as the cyclic middle fits into a cavity formed by hydrophobic amino acids.

The binding cavity of the oestrogen receptor is both rigid and flexible. Rigid regions mediate recognition of the phenolic component of the ligand, as well as the separation of the hydroxyl groups. Flexible regions accommodate extension of the ligand core, for example, inclusion of antagonistic side chains, and permits alternate binding modes of the spacer component. The key binding residues of the oestrogen receptor are shown in Figure 7.5. Remember that drug efficacy relates to biological activity, not strength of ligand binding. The structural basis of agonism/antagonism relates to competition for the surface responsible for co-regulator binding: different ligand classes induce or stabilise the distinct orientation of helix-12, a region of the oestrogen receptor that dictates the formation of the co-regulator binding cleft. In this way, oestrogen analogues can be designed with different side chains to achieve different biological activity. This illustrates how rational drug design can be adopted to produce novel therapies to battle diseases, in this case, cancer.

7.3 FRONTIERS OF CANCER THERAPY

Target specificity is crucial to the success of cancer therapy. This is because the correct dosing of the drug to achieve therapeutic concentrations at the site of the cancer, i.e. the tumour, is required for the effective remediation of the tumour as well as minimising the harmful side effects that are synonymous with traditional chemotherapy. Another strategy to battle cancer is to target the structural elements of the cell division process. Disruption of mitosis will obviously impede cell proliferation. Tubulin is a structural protein that is crucial for cell division, involved in polymerisation and depolymerisation of microtubules, where it is used as protein building blocks. As the cells divide, mitotic spindle fibres link the two daughter cells. The spindle fibres are made from microtubule polymers formed from tubulin proteins. Drugs designed to bind to tubulin prevent its polymerisation to from the microtubules, and thus cell division is arrested. Delivering the chemotherapy to the correct location is therefore highly desirable for cancer therapy. Utilising a pro-drug approach with antibodies is another way in which this can be done.

7.3.1 ANTIBODY-DIRECTED THERAPIES

Antibody-directed therapies take advantage of the unusual morphology of cancer cells, where the altered plasma membrane contains distinctive antigens, which are over-expressed. These antigens are much more numerous on cancer cells than normal cells and monoclonal antibodies can be utilised as anticancer agents, with the desired mode of action being to activate the body's immune system to direct killer cells towards the tumour. The antibody can also act as a receptor antagonist.

The activity of monoclonal antibodies in this instance is quite low. An improved strategy is to attach an anticancer drug to the antibody in the form of an antibody–drug conjugate, which can facilitate the selective delivery of the anticancer drug to specific types of cancer cells. This selectivity is highly desirable because, for many anticancer drugs, the concentration needed for the effective treatment of cancer is close to the threshold concentration for toxicity.

The first-generation antibody–drug conjugates used anticancer drugs such as methotrexate and doxorubicin but yielded disappointing results with lower anticancer activity and the same toxicity compared to administration without the antibody as a vector. This is because the lifetime of the antibody–drug conjugate was much longer than that of the free drug, which was a contributing factor to the toxicity problem. Also, the size of the antibody impeded penetration into the tumour and limited the amount of drug being delivered, hence reduced activity. It was realised that more potent anticancer drugs were needed for application in antibody-directed therapy. These drugs would require a stable bond to the antibody, so that it remains bound until reaching the tumour and not release the drug into the bloodstream, which would result in high toxicity.

A drug can be bound to an antibody in a number of ways. Lysine residues are present in many parts of the antibody molecule and contain a nucleophilic primary amine functional group onto which a drug molecule could be added by simple acylation or alkylation reaction. However, care must be taken not to attach the drug to groups present in the binding region, or recognition site, for the cancer antigen. The drug must not obstruct antigen–antibody binding. To avoid obscuring the antibody binding surface, a good approach is to reduce the four intra-strand disulphide bridges at the hinge of the Y-shaped antibody protein molecule to produce eight thiol groups to which a drug may be added by alkylation. While this approach limits a maximum of only eight drugs being attached, a linker molecule can be attached on to the antibody, which can contain several drug molecules. Alternatively, the carbohydrate region of an antibody can be utilised for drug attachment by lightly oxidising the vicinal diols of the sugar rings to produce aldehyde groups onto which drug molecules can be linked.

This linker must be cleaved only when the drug–antibody conjugate has entered the cancer cell, otherwise the method will lose its effectiveness. For example, a disulphide linker may be cleaved by disulphide exchange with an intramolecular thiol, such as glutathione, which has a higher concentration inside the cancer cell than in plasma. It is crucial that the drug is only cleaved inside the cell because drugs used in this application are highly potent, with IC_{50} values $<10^{-10}$ mol dm^{-3}, which are 100–1000 times more cytotoxic than conventional anticancer drugs.

Antibody-directed enzyme prodrug therapy (ADEPT) is another method utilising the selectivity of antibodies. In this instance, an antibody–enzyme complex is administered, where the antibody, which is linked to a bacterial enzyme, binds selectively to tumour antigens, but remains attached to the surface of the cell and must not be internalised, in contrast to antibody–drug conjugate therapy. After sufficient time has elapsed for all the antibody–enzyme complex to become bound to its target antigen, or expelled from the bloodstream, a cytotoxic prodrug is administered. The design of these drugs ensures stability in the bloodstream, and they can only be cleaved to reveal the active compound by the enzyme complexed to the antibody, thus ensuring that the highly cytotoxic drug is concentrated in the tumour and the effect on normal cells is minimal.

Cleavage of the active cytotoxic drug occurs by enzymatic breakdown of the β-lactam moiety. A nucleophilic group in the active site attacks the unstable β-lactam at the electrophilic carbonyl carbon resulting in ring-opening, while the electrons from the bond resonate along the molecular structure to cause two further bond cleavages and release the active compound. This mechanism is shown in Figure 7.6. Because the enzyme is anchored to the antibody, which selectively binds to cancer cells, the prodrug only encounters the enzyme within the cancerous tissue, so this is the only location where the active compound is unveiled, thus minimising side effects.

This method has advantages over antibody–drug therapy. Enzymes are catalytic, so can generate a large number of active drug molecules at the tumour site, which diffuse into the tumour and may also target cancer cells in this region that do not display the necessary antigens for antibody binding. The use of foreign enzymes derived from bacteria circumvents the problems that would arise if a mammalian enzyme

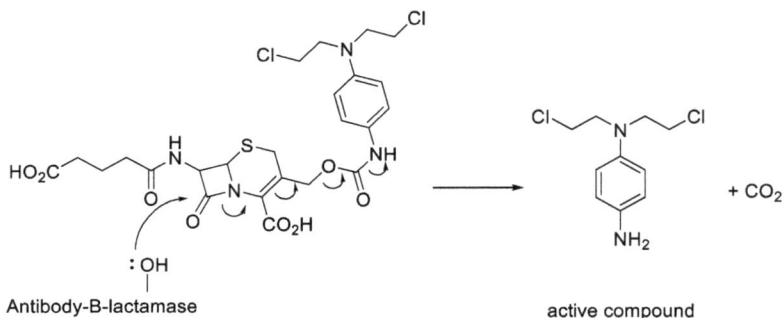

FIGURE 7.6 Mechanism for the enzymatic cleavage of a prodrug to produce the active compound for the ADEPT strategy to tackle cancer.

was to be used, which could result in similar enzymes within the body activating the highly cytotoxic prodrug in healthy tissue.

Complications could arise from the use of foreign enzymes in ADEPT in regard to immune responses. The use of human enzymes, as mentioned before, carries the risk of the prodrug being activated elsewhere in the body. Also, the activity of human enzymes may be lower than bacterial enzymes, so lower levels of drug would be present in the tumour. However, research into gene therapy has produced techniques that could rectify these problems.

7.3.2 Gene-Directed Approaches to Cancer Treatment

Gene-directed enzyme prodrug therapy (GDEPT) uses implanted genes into the cancer cell's DNA, which codes for the enzyme used to activate the prodrug. This is achieved through use of a vector, such as a retrovirus or adenovirus, where the desired gene is spliced into the viral DNA, and so is inserted into the host cancer cell on infection. Additionally, the virus is modified so that it is no longer virulent and cannot harm normal cells. Typically, the vector is administered directly into the tumour for greatest effect.

Once the necessary gene has been inserted, the cell will start to produce the desired enzyme. Because the foreign enzyme is produced inside the cell, it will not be subjected to immune responses, so this technique circumvents this issue that had important consequences for ADEPT. However, it is unlikely that the desired gene will be delivered to all the cancer cells, so it is important that the drug be transferred between the cancer cells.

Further gene-directed research is at the forefront of modern cancer research. By understanding the molecular basis of this complex disease unique and unequivocal targets can be identified for which treatments can be developed that have much lower side effects compared to the traditional anticancer therapies. Cancer is one of the main challenges to modern medicine, where prevalence among an ageing population has increased, and the consequences of the disease are devastating. The pioneering research done by medicinal chemists though offers a chance for us to overcome this disease and perhaps create a future where the threat of cancer in old age is less ominous.

TUTORIAL QUESTIONS

1. State two requirements for a drug to be used in ADEPT.
2. The nitrogen mustard prodrug below is used in Antibody-Directed Prodrug Therapy (ADEPT) with a bacterial β-lactamase enzyme. With the aid of a curly arrow mechanism, explain how the nitrogen mustard is released on enzyme activation of this compound.

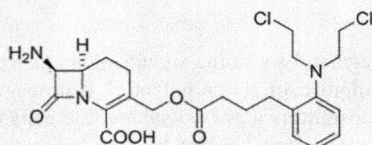

3. Describe two advantages of using a prodrug approach, such the β-lactam attached to the mustard compound, as a cancer therapeutic in comparison to the simple nitrogen mustard.
4. 17-β-Oestradiol (left) and benzopyran (right), shown below, act at the breast oestrogen receptor. Identify the key structural features that enable each to act at the breast oestrogen receptor and describe the mode of action in each case.

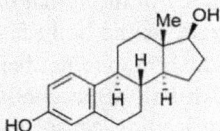

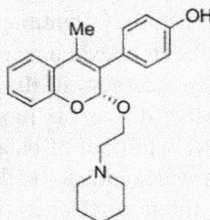

5. Tegafur is an irreversible inhibitor of the enzyme thymidylate synthase. Explain what is meant by the term 'irreversible inhibitor'. Hence, discuss how tegafur inhibits thymidylate synthase and can therefore be used as a cancer therapeutic.

8 Drugs to Keep the Body in Balance

ABSTRACT

The nerve impulse is important for sending signals along neurons between the brain and the body. The nerve impulse, or action potential, is propagated between neurons by the passage of neurotransmitters across synapses, the gaps between neurons. For example, the neurotransmitter acetylcholine operates within cholinergic receptors and has an effect on heart rate and smooth muscles such as those in the intestine. In addition to the nervous system, the body is also kept in balance by hormones in the endocrine system. For example, adrenaline has an influence on heart rate in conjunction with other hormones and neurotransmitters. Insulin, for example, is important in balancing and regulating blood sugar levels, and where derangements occur in cell's receptivity to insulin, diabetes may occur. Medical approaches need to be implemented to bring about balance to the body to counteract such diseases as diabetes and heart disease.

A disease may be defined as a malfunction of the body or mind that has an adverse effect on good health. Oftentimes, diseases can arise, as the body falls out of balance. Examples can be when neurotransmitters in the brain are not being produced, or registered by receptors in the normal way, which will affect mood. In old age, neurodegenerative disease is more common and a leading cause of death. When blood sugar levels fall out of balance, this can lead to diabetes, which is becoming increasingly prevalent due to modern, sugar-rich diets, as well as other factors. Heart disease, when the normal operation of that most important of muscles falls out of rhythm, likewise is becoming increasingly common and is among the highest causes of death. With the prevalence of diseases occurring as people's bodies fall out of balance, there is a greater incentive for pharmaceutical companies to develop drugs to restore balance to the body. As such, Alzheimer's, dementia, diabetes and heart disease are major earners for big pharma and therefore a very active area of research and development.

8.1 DRUGS TARGETING THE NERVOUS SYSTEM

A thorough understanding of the nervous system, the role of neurotransmitters and hormones in the endocrine system, is essential for understanding the molecular basis of diseases that arise from malfunctions in these systems and for identifying targets that drugs can be developed for as a means to cure these diseases affecting the nervous system.

DOI: 10.1201/9781003516835-8

8.1.1 THE NERVE IMPULSE

The nervous system is composed of two parts: the central nervous system (brain and spinal cord) and peripheral nervous system, which extends to the entire body. Sensory neurons carry information from the body to the CNS, while motor nerves carry messages from the CNS to the rest of the body. Information from a stimulus is carried to the CNS by the sensory neurons; here the messages are coordinated, and the appropriate messages are then sent from the CNS to effectors, which can be organs or muscles, to generate a response to the stimulus. For example, if a person was to touch a hot object, for example, they accidentally touch the ring on a hot cooker, the information from the stimulus, measured by temperature receptors in the skin, travels via a sensory neuron to a coordinator, such as a connector neuron within the CNS, which produces an automatic response to move the hand away from the heat by sending signals down motor neurons to the effector muscles. This pathway of neurons is known as a reflex arc. The response is immediate because only three neurons are involved, and it is an automatic, involuntary response because it bypasses the brain to save time and avoid damage.

These signals can be considered as electrical pulses, but rely on the movement of ions, rather than electrons. As the ions move across the membrane of a neuron (nerve cell), a potential difference is created, which can be reversed to propagate an electrical signal: a nervous transmission, known as an action potential. Nerves are composed of many tightly packed neurons forming a bundle held together by connective tissue and have a blood supply. They are distributed throughout the body like cables and run in distinct pathways. A nerve may contain both sensory and motor neurons, but the proportions of each can vary.

A nerve impulse is a self-propagating wave of electrical disturbance, not current, that travels along the axon membrane of a neuron, due to a temporary reversal of electrical potential difference. This reversal alternates between two states: the resting potential and an action potential. There is a threshold level of stimulus needed to propagate an action potential; this feature is referred to as the 'all or nothing principle'. The magnitude of a stimulus is perceived by the number of impulses passing in a given time and/or having neurons with different threshold values, which is left for the brain to interpret.

A resting potential of around -65 mV is maintained by the active transport of Na^+ ions out of the nerve axon and K^+ ions into the axon via intrinsic protein channels that make up the 'sodium–potassium pump'. Three sodium ions are pumped out for every two potassium ions that enter the axon. Meanwhile, K^+ ions are able to diffuse back out of the axon through specific ion channels, while sodium ion channels are mostly closed. This creates an electrochemical gradient; tissue fluid outside the neuron is positive compared to the axon membrane, which is said to be polarised.

The energy from a stimulus causes a temporary reversal of charge in the axon membrane, due to the opening of voltage-gated channels. As the action potential passes, the sodium–potassium pump is turned off, while sodium voltage-gated channels open, allowing Na^+ to diffuse into the axon along their electrochemical gradient and potassium ion channels close. This causes a reversal of charge across the axon membrane and an action potential of $+40$ mV is generated. This wave of depolarisation proceeds along regions of the axon, causing sodium ion channels to open, and then close in a Mexican wave-like fashion. These channels then remain closed during a refractory period to ensure that the nerve impulses are discrete and unidirectional. Repolarisation occurs when the potassium ion channels open again and K^+ diffuse back out of the axon and the sodium–potassium pump reactivates.

A nerve route may comprise several different neurons. The point at which two neurons connect is called a synapse. Synapses act as junctions when nerve impulses are being transmitted, allowing a single impulse along one neuron to be transmitted along a number of different neurons at a synapse, allowing a single stimulus to generate a number of simultaneous responses. Equally a number of impulses can be combined at a synapse, which allows one response to be generated for stimuli from different receptors.

8.1.2 SYNAPSES

Where two neurons meet at a synapse, the nerve impulse is interrupted because there is a gap between the neurons, called the synaptic cleft. In order for the signal travelling down the presynaptic neuron to be continued, it must be propagated in the postsynaptic neuron across the synapse via neurotransmitters, such as acetylcholine. Neurotransmitters are molecules that, upon activation by an action potential, are released from the presynaptic neuron into the synaptic cleft. The neurotransmitter molecules diffuse across the gap and bind to receptors on the postsynaptic neuron, causing a chain of reactions that result in the propagation of the action potential in the postsynaptic nerve and the signal can proceed to its final destination. The actions of the neurotransmitter acetylcholine are instrumental in the operation of the cholinergic nervous system.

Acetylcholine is synthesised from choline and acetyl coenzyme A in the nerve ending of the presynaptic nerve, catalysed by the enzyme choline acetyltransferase. Acetylcholine is incorporated into membrane-bound vesicles by means of a specific carrier protein. An action potential results in the opening of calcium ion channels and increases the intracellular Ca^{2+} concentration, which induces the vesicles to fuse with the cell membrane of the presynaptic nerve and release acetylcholine in to the synaptic cleft. The neurotransmitter crosses the synaptic cleft and binds to cholinergic receptors on the surface of the postsynaptic neuron. This has the effect of opening sodium ion channels in the postsynaptic membrane and propagating the action potential in the postsynaptic nerve, whereby the signal can carry on to its destination. The enzyme acetylcholinesterase hydrolyses acetylcholine into choline and ethanoic acid, which diffuse back into the presynaptic neuron. This breakdown prevents new action potentials from being continuously generated.

8.1.3 THE PERIPHERAL NERVOUS SYSTEM

Nerves transmit signals between the CNS, made up of the brain and spinal cord, to the body (peripheral nervous system). The motor nerves of the **peripheral nervous system** are divided into three subsystems. The somatic motor nervous system carries signals from the CNS to skeletal muscle to stimulate voluntary muscle contraction.

The autonomic nervous system carries messages from the CNS to smooth muscle, cardiac muscle and the adrenal medulla to stimulate the release of adrenaline. These nervous impulses are divided across two pathways. Parasympathetic nerves leave the CNS, travel some distance before encountering a synapse with a second nerve, then the nerve impulse is transmitted across the junction using the neurotransmitter acetylcholine. Sympathetic nerves leave the CNS, but almost immediately synapse with a second nerve, again using acetylcholine. The second nerve proceeds to the same target organs as the parasympathetic pathway, but the synapses also have different receptors that use a different neurotransmitter: noradrenaline. These two systems tend to have an antagonistic relationship, where the sympathetic nervous system gets the body ready for action, and the parasympathetic calms the body down.

The enteric nervous system is located in the walls of the intestine and receives messages from parasympathetic and sympathetic nerves as well as responding to local reflex pathways, involving a variety of neurotransmitters. The enteric nervous system tends to act more independently and is occupied with habitual modulation of the gastrointestinal system.

The sympathetic system promotes the 'flight or fight' response, where noradrenaline is released and promotes contraction of cardiac muscle, associated with increased heart rate, and relaxes smooth muscle, reducing contraction of GI tract and urinary tract. The elevated heart rate and suppression of general bodily operations prepare the body for action. Moreover, stimulation of the adrenal medulla releases the hormone adrenaline, which reinforces the process. The parasympathetic pathway leads to opposite effects: acetylcholine is released to target organs and stimulates bodily processes.

These opposing systems are kept in balance in the body, where the proportionate strength of each response depends on physical stimuli. It is crucial for a body to remain in homeostasis in order to function properly; failure in either of these systems could lead to ailment in the heart, skeletal muscle or digestion, which could be a consequence of deficiency or excess of neurotransmitter, hence the potential for developing agonists/antagonists to serve as treatments for conditions that affect these organs.

Consider the processes taking place in the cholinergic signalling system, as messages are transmitted between neurons or from a neuron to an organ, such as a muscle, across a synapse, using the neurotransmitter acetylcholine. The neurotransmitter is released from the presynaptic neuron during an action potential. The acetylcholine molecules travel across the synaptic cleft and must bind to cholinergic receptors in the membrane of the postsynaptic neuron. This then leads to processes that propagate an action potential in this nerve. The cholinergic receptor therefore presents an attractive target for drugs that can be designed to influence the nervous system.

8.1.4   THE CHOLINERGIC RECEPTOR

The purpose of **cholinergic receptors** (autoreceptors) present at the terminal of the presynaptic nerve is to provide a means of local control over nerve transmission. The binding of acetylcholine to the autoreceptor has the effect of inhibiting further release of acetylcholine. Also, the presynaptic nerve contains noradrenaline, which serves as another system of presynaptic control of acetylcholine release. Noradrenaline is released when the sympathetic nervous system is active and binds to those receptors in the cholinergic synapse and has the effect of inhibiting the release of acetylcholine. If derangements occur, where there is an insufficient release of acetylcholine, it would be logical for medicinal chemists to develop an agonist for the cholinergic receptor. While acetylcholine can be readily synthesised in the laboratory, it is not feasible as a treatment due to being hydrolysed in the stomach and blood and with no selectivity of action. Hence analogues of acetylcholine must be developed that are stable to hydrolysis and selective where they act in the body.

Selectivity can be achieved through designing synthetic analogues with subtle structural differences that result in optimal binding with a particular cholinergic receptor at one part of the body over another. This is possible as slight differences in the structures of these receptors are observed at different parts of the body. Also, the drug might be preferentially distributed to one part of the body over another. There are subtle differences in cholinergic receptors around the body, for example, although the binding region for acetylcholine is conserved, there may be a molecular barrier present on the periphery that might block access to a larger synthetic analogue. This is also true for other types of receptors.

Differentiation among cholinergic receptors was first observed from studies into the physiological effect of nicotine (present in tobacco) and muscarine (the toxic compound in poisonous mushrooms). These are both acetylcholine agonists, but nicotine was found to be active between different nerves at the synapses between skeletal muscle and its motor units but had poor activity elsewhere. Muscarine, by contrast, had poor activity in these places, and was found to affect synapses of nerves with smooth muscle and cardiac muscle. It was concluded that these two types of receptors, nicotinic and muscarinic receptors, had structural differences and demonstrated that receptor selectivity was possible. However, now there was the challenge to develop drugs without side effects.

This led to the study of structure–activity relationships (SAR) of acetylcholine to understand receptor binding and elucidate which aspects of the compound could be modified for selectivity. The positively charged nitrogen atom is essential, as is the distance from the atom to the ester group, so the overall size of the molecule cannot be altered significantly and the ester bridge must be important and therefore has to remain intact. Once the essential features of the acetylcholine–receptor complex were identified and understood, medicinal chemists could work to develop compounds that would successfully bind with a particular cholinergic receptor but have subtle structural differences that would give these compounds receptor selectivity.

The conclusions drawn from this research are that acetylcholine fits tightly in its binding site, so there is little opportunity for variation. Important hydrogen bonding interactions presumably exist between the ester group of acetylcholine and an asparagine residue, while a small hydrophobic pocket is present into which the methyl ester can fit, but nothing larger. From the evidence, it appears that the NMe_3^+ group is placed in a hydrophobic pocket lined with aromatic amino acids, made of three compartments, two of which are only large enough to accommodate methyl groups, while the third has enough space to tolerate a larger substituent. It is proposed that a strong ionic interaction exists between the positive nitrogen and an anionic aspartic acid residue. An alternative suggestion involves an induced dipole interaction between the

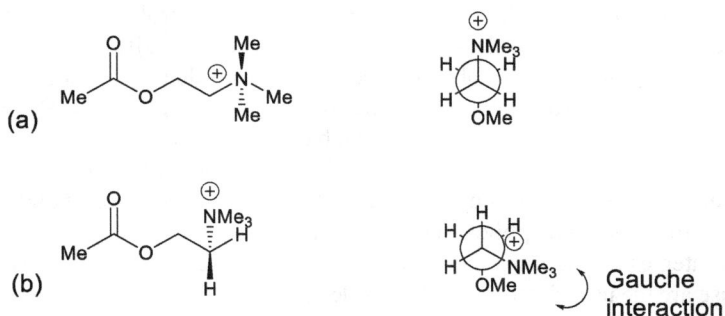

FIGURE 8.1 Molecular structure of acetylcholine drawn as the staggered conformation (a) and as the gauche conformation (b); viewing the molecule down the ester group.

NMe_{3+} and aromatic amino acids in the hydrophobic pocket, based on the nature of the positive charge being delocalised onto the methyl groups, and therefore are less likely to be involved in localised ionic bonding.

Identifying the active conformation of acetylcholine was important for understanding its binding. Acetylcholine is highly flexible; bond rotation along the length of its molecular framework can produce at least nine possible stable conformations, and medicinal chemists strived to determine which of these shapes is exhibited in the neurotransmitter–receptor complex. This would enable design of analogues with the appropriate shape for binding.

Initially, it was presumed that acetylcholine would adopt its most stable conformation, illustrated in Figure 8.1(a) which shows sawhorse and Newman projections at the energy minimum. However, this was inaccurate: insignificant energy differences between the alternative stable conformations, such as the gauche conformation, Figure 8.1 (b), means that stabilisation energy gained from binding interactions with the receptor can more than compensate for discrepancies in conformational energies.

The illustrations in Figure 8.1 show the stereochemistry of acetylcholine: rotation about the bonds means that the amine (NMe_3) can be located at different positions relative to the rigid C=O double bond. These conformations are associated with different energies, due to steric effects that arise as the amine NMe_3 and ester OMe groups occupy the same space while in the gauche conformation; as can be clearly seen when the molecule is viewed down the ester group in the Newman projections on the right-hand side of Figure 8.1.

To elucidate the active conformation of acetylcholine, rigid cyclic molecules were studied which contained the molecular framework of acetylcholine, except that the conformations in these analogues are fixed, due to the rigid cyclic component in which the bonds cannot rotate freely. If such a molecule could bind to the receptor, it demonstrates that this particular conformation is 'allowed' for activity. By understanding the binding interactions and the nature of the pharmacophore of acetylcholine, medicinal chemists were able to develop clinically useful analogues of acetylcholine.

As previously mentioned, acetylcholine is prone to hydrolysis. This instability is an issue that had to be addressed in order to make clinically useful analogues. The primary reason for the instability of acetylcholine is due to neighbouring group participation of the nitrogen atom, making the carbonyl carbon more electrophilic. In a particular conformation, the positively charged nitrogen interacts with the carbonyl and has an electron-withdrawing effect. To compensate, the oxygen atom pulls electron density from the carbonyl carbon, making it electron deficient and therefore susceptible to weak nucleophiles, such as water.

Two strategies can be implemented to tackle the inherent instability of acetylcholine. These are the use of steric shields and electronic stabilisation. The use of a steric shield involves incorporating a bulky substituent on the ethyl bridge to protect the ester group. It does this by causing steric hindrance to the approaching nucleophiles that would otherwise initiate a hydrolysis reaction. It also interrupts the binding of esterase to inhibit enzymatic hydrolysis. The only position available for attaching a steric shield is on the ethyl bridge because, as previously mentioned, the tight fitting of acetylcholine in its receptor prohibits too much structural variation. Moreover, the inclusion of an extra methyl group to serve as a steric shield is the only option, as larger groups are not tolerated.

Electronic properties can also be used to stabilise functional groups. In the case of acetylcholine analogues, the methyl ester is converted into a carbamate to give carbachol. The lone pair of electrons on the nitrogen of the NH_2 group can be donated into the carbonyl group and therefore reduce its electrophilic character and stabilise the compound against hydrolysis. Carbachol shows good stability to chemical and enzymatic hydrolysis but does not show selectivity between the two receptor types. Nevertheless, it was a useful treatment for glaucoma, where it could be applied locally, circumventing the issue of receptor selectivity. Glaucoma is caused by pressure building in the eye when its aqueous contents cannot be drained properly and can lead to blindness. Agonists cause the muscles in the eye to contract and relieve the blockage and carbachol proved a useful treatment for this condition.

The stability of the carbamate functional group and the receptor selectivity provided by the β-methyl group meant that medicinal chemists could combine steric and electronic properties to produce a stable, selective muscarinic agonist and this led to the synthesis of bethanechol, which is used therapeutically to stimulate the GIT and urinary tract after surgery.

Antagonists of the muscarinic cholinergic receptor are also therapeutically important. They work by binding to the receptor without initiating a biological response, while blocking access to the receptor by the neurotransmitter, acetylcholine. In this way, the usual biological process is suppressed. The clinical uses of muscarinic antagonists are for shutting down the GIT and urinary tract during surgery by relaxing smooth muscle. They are also useful for dilation of pupils for ophthalmic examinations and reducing glandular secretions.

The first antagonists to be discovered were natural products extracted from plants; in particular, nitrogen-containing compounds called alkaloids. For example, atropine is derived from the roots of *Atropa belladonna* (deadly nightshade) and was used in the past by Italian women to dilate their pupils, so that they would appear more beautiful, which is where the name 'belladonna' comes from. Atropine has been used clinically to reduce gastrointestinal motility. As typical for most natural products, atropine exists as one enantiomer, but racemisation occurs when it is extracted into solution. This compound binds to the cholinergic receptor on account of a charged nitrogen atom when protonated, and an ester group at the required distance apart. Atropine acts as an antagonist because the molecule is larger than acetylcholine and has different binding interactions with the receptor, so it doesn't induce the same conformational change on the receptor as acetylcholine, and the receptor is not activated.

Atropine is a tertiary amine rather than a quaternary salt and so can pass the blood–brain barrier as the free base, become protonated and then antagonise muscarinic receptors in the CNS. This can cause hallucinogenic activity and restlessness. To reduce side effects on the CNS, structural analogues based on atropine were developed, such as quaternary salts of atropine, which are used clinically.

Antagonists of the nicotinic cholinergic receptor are also medicinally beneficial. They are present in nerve synapses at neuromuscular junctions and can be used as blocking agents. In the 16th century, when Spanish soldiers, known as conquistadors, invaded South America, the indigenous people used poisonous arrows in retaliation. The Indians used a crude dried extract from a plant called *Chondrodendron tomentosum*, which caused paralysis and stopped the heart. This extract is known as curare, consisting of a mixture of compounds, but later discovered that the active component was an antagonist of acetylcholine, which blocks nerve transmissions from nerve to muscle. This compound is called tubocurarine, and with controlled dose levels is found to be medicinally useful for relaxing the abdominal muscles before surgery; therefore, a lower dose of general anaesthetic can be used, improving safety of operations. However, side effects on the autonomous nervous system meant that better drugs had to be developed.

The design of the drug atracurium was based on the structure of tubocurarine. This is an improved therapeutic agent, since it is free from cardiac side effects and is rapidly broken down in the bloodstream, so it can be administered as an intravenous drip. The drug is rationally designed to break down via a Hofmann elimination reaction, which proceeds at blood pH. The positive charge on the nitrogen is lost and the molecule splits in two. This is ingenious because the particular feature of the compound that gives its biological activity promotes it deactivation.

The important structural features of the drug are the spacer: a 13-atom chain which connects the tertiary centres, the blocking units, which are cyclic structures at either end of the molecule to block access of acetylcholine, and the quaternary centres which are essential for receptor binding. If one of these centres is lost through a Hofmann elimination, the binding interaction becomes too weak, and the molecules evacuate the receptor. The Hofmann elimination is facilitated by the ester group, which serves as an electron-withdrawing group on the carbon β to the quaternary nitrogen atom, therefore increasing acidity of the H on the β carbon, easily losing in weakly alkaline conditions and the drug becomes deactivated. As a consequence, the drug acts briefly, about 30 minutes, so is administered as an IV drip for the duration of surgery. Once the drip is removed, antagonism ceases almost immediately.

There are other avenues available to target the cholinergic receptor. Once acetylcholine has stimulated a neuron, it must be hydrolysed in order to prevent reactivation of the cholinergic receptor, which would increase cholinergic effects. This destruction of acetylcholine is done by an enzyme called acetylcholinesterase. Inhibitors of this enzyme, referred to as anticholinesterases, will have the same biological effect as a cholinergic agonist.

To be able to design anticholinesterases, it is necessary to understand the active site of acetylcholinesterase, and the binding interactions with acetylcholine, along with the mechanism of hydrolysis. There are two features of the active site that are crucial for substrate binding. These are intermolecular forces and positioning of catalytic residues. Ionic interactions with an aspartate residue and hydrogen bonding to a tyrosine residue form during substrate binding, and an aspartate, histidine and serine residues are involved in the mechanism of hydrolysis.

Anticholinesterase drugs inhibit the active site of acetylcholinesterase reversibly or irreversibly, depending on the interactions with the active site. The two main groups of anticholinesterases include carbamates and organophosphorus compounds. The lead compound for the carbamate inhibitors was sourced from the natural product physostigmine, which was discovered in 1864 as a product from the poisonous calabar bean from west Africa; its structure was determined in 1925. This compound is still used clinically to treat glaucoma. SAR studies show that the carbamate group is essential to the activity, the benzene ring is important and the pyridine nitrogen is ionised at blood pH, crucial for binding to anionic residues in the active site. The carbamate group is crucial for the inhibitory properties of physostigmine. The mechanism for hydrolysis produces a stable carbonyl intermediate which is the rate-determining step. Molecular structures for some of these compounds are given in Figure 8.2. Due to serious side effects, its medical uses are limited, so analogues have been made that retain these important features.

8.1.5 ADRENALINE AND THE PERIPHERAL NERVOUS SYSTEM

In addition to the cholinergic systems that rely on the neurotransmitter acetylcholine, the peripheral nervous system also includes the adrenergic system, which uses adrenaline and noradrenaline as chemical messages. The neurotransmitter noradrenaline is released by the sympathetic nerves, which supply smooth muscle and cardiac muscle with stimulus. Adrenaline is a hormone released along with noradrenaline from the adrenal medulla located in the kidneys and circulates in the blood supply in order to reach remote organs. Activation of adrenergic receptors promotes physiological effects, such as increasing heart rate and expanding lung capacity.

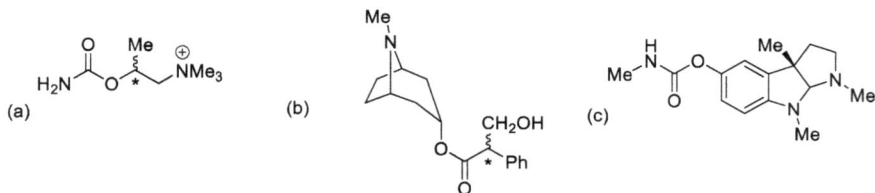

FIGURE 8.2   Molecular structures of bethanechol (a), atropine (b) and carbamate (c).

Tissues tend to be under a dual control system, where noradrenaline has the opposite effect to acetylcholine. Both the cholinergic system and the adrenergic system have a background level of activity and the overall response depends on which stimulus is predominant. An additional feature of the adrenergic system is the facility to release the hormone adrenaline during the 'fight or flight' response to prepare the body for immediate action by stimulating the heart and dilating the blood vessels to the muscles in times of perceived danger or stress, while shutting down digestion. The CNS also contains adrenergic receptors; noradrenaline has an important role in sleep, emotion, temperature regulation and appetite.

There are two different types of adrenergic receptors, known as α- and β-adrenergic receptors, which are G-protein-coupled receptors, differing in the type of protein that they are coupled to. Moreover, these two receptors have different sub-types with a different distribution around the body and have subtle structural differences that can allow for selective drug design to target specific organs. The neurotransmitter noradrenaline and hormone adrenaline both activate the adrenergic receptors and belong to a group of compounds called catecholamines. These compounds consist of a catechol ring (1,4-benzenediol) linked to an alkyl amine chain and are synthesised from the amino acid tyrosine. They are metabolised via enzymatic pathways.

Binding to the adrenergic receptor involves the important function groups on catecholamine. Structure–activity relationships show the importance of the alcohol groups and the phenol catechol ring, as well as the ionised amine needed for binding roles in the receptor. The alcohol group is involved in hydrogen bonding; indicated by the R-enantiomer of noradrenaline being more active than the S-enantiomer. Compounds without the secondary alcohol group, e.g. dopamine, have greatly reduced activity, showing that the alcohol is important but not essential for binding. The amine is normally ionised, being protonated at physiological pH, needed for ionic bonds to the Asp-113 anion in the receptor. Activity drops with substituents on the tertiary amine/quaternary salt.

SAR studies also demonstrate features that introduce a level of selectivity between the α- and β-receptors. Adrenaline has the same potency for each type of receptor, but noradrenaline has greater affinity for α-receptors, which suggests that N-alkylation has implications for receptor selectivity. Indeed, increasing the size of the N-alkyl substituent results in an increase in activity for the α-receptor and a decrease in potency for the β-receptor. This is due to structural differences in the receptor binding sites: the α-receptor contains a hydrophobic pocket into which the alkyl substituent can fit, whereas this pocket is absent in the β-receptor binding site. Furthermore, extension of the N-alkylated substituent to include a polar functional group introduces favourable binding interactions, resulting in a dramatic increase in activity. For example, this can be achieved by adding a phenol group to the alkylated amine. This led to the development of adrenergic agonists, with selectivity.

The treatment of asthma demonstrates the usefulness of adrenergic agonists. Drugs specifically designed to target β$_2$-receptors can be used to relax smooth muscle. In the treatment of asthma, this causes dilation of the airways, and because β$_2$-adrenoreceptors predominate in the bronchioles, this allows for selectivity. Originally, adrenaline was used to treat asthma attacks, but because there was no selectivity for β$_2$-receptors, it caused stimulation of adrenergic receptors around the body and particularly had cardiovascular side effects. Elevated heart rate increased the body's demand for oxygen which counteracts the effects of opening the airways. The use of adrenaline is restricted to emergencies now.

The development of isoprenaline, where a bulky isopropyl substituent is added to the nitrogen, provided selectivity towards β-receptors over α-receptors, but still stimulated β_1-receptors in the heart and the problems with short duration of activity continued. Metabolic enzymes rapidly degrade these drugs to form an inactive ether. To make the drugs more resistant to metabolism, attempts were made to modify the meta phenol group, so that the drug would not be recognised by the enzymes. Due to the importance of this group to activity, the modifications that were made needed to preserve the hydrogen bonding interactions with the receptor binding site in order to retain biological activity. A variety of different substituents were tested to understand the binding role of the phenol, for example, the size of the group and electronic effects were varied and screened. Carboxylic acid groups had no activity and esters/ amides were β-receptor antagonists. This research led to the discovery of salbutamol, where the hydroxymethylene group is used. The similarity between these structures is illustrated in Figure 8.3.

Salbutamol has the same potency as isoprenaline, but is 2000 times less active on the heart, and with a duration of action of 4 hour, not being recognised by certain metabolic enzymes. Salbutamol became a market leader for treating asthma in 26 countries. It was marketed as a racemate, but the R-enantiomer is 68 times more active than the S-enantiomer, which also accumulates in the body to give side effects. As a result, the R-enantiomer (levalbuterol) was marketed.

Salbutamol, patented in 1969, is a great example of drug design in action. The main problem with its predecessors was that metabolism was limiting duration of action. It was discovered that the OH group in the 3-position becomes methylated during metabolism and activity is lost as a consequence. This therefore may be regarded as a pharmacokinetic problem. By replacing the OH group with a CH_2OH group, binding interactions were preserved, but importantly metabolism was much slower because the terminal hydrogen of CH_2OH is less acidic, increasing the duration of action to 4 hours. This revolutionised the treatment of the disease, which has greater than 150 million sufferers worldwide.

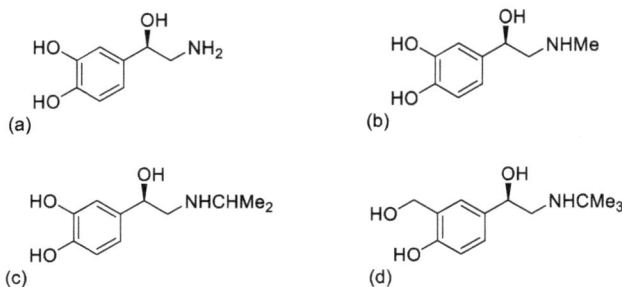

FIGURE 8.3 Structures of (a) noradrenaline, (b) adrenaline, (c) isoprenaline and (d) salbutamol.

FIGURE 8.4   Structures of the neurotransmitters (a) dopamine and (b) serotonin.

8.1.6 DRUGS OF ABUSE

The hormone adrenaline is closely related to a range of vital neurotransmitters, each of which has an important role in determining mood and behaviour. Intriguingly, all of these neurotransmitters are structurally related, with only subtle differences to optimise their binding to specific receptors. Noradrenaline is the most closely related to adrenaline and is responsible for the amount of stimulation experienced. Dopamine is associated with the sense of 'reward' when performing a task, and serotonin is involved in mediating the biochemical processes of feeling good: so-called 'euphoria'. In order to carry out their roles in the brain, these neurotransmitters must dock with key enzymes, which catalyse the biochemical processes associated with mood and behaviour. The structures of the neurotransmitters dopamine and serotonin are shown in Figure 8.4. Structural analysis and amino-carboxylate interactions, in combination with hydrophobic interactions with the aromatic rings, are vital for the general mode of action of neurotransmitters in conjunction with enzyme binding. Other aspects of the molecular structure can be subtly varied to tune the precise activity profile. This means that this family of compounds can easily be used to control mood.

Stimulant and mind-altering drugs of this kind have been used in different forms throughout history. Mescaline, found naturally in the peyote cactus, is used by native North Americans to access a pseudo-religious trance. Another psychoactive plant extract belonging to this class of compounds is ephedrine, known since ancient times in China. The first fully synthetic 'recreational drug' is an analogue of ephedrine, amphetamine, first synthesised in 1887. It has been used to cheat in sports, to increase heart rate and circulation to improve the performance of the athlete, but this was risky and resulted in deaths. Interestingly, pseudoephedrine (a diastereomer of ephedrine) is used in many over-the-counter cold remedies but has a much lower stimulant effect than ephedrine. Even though it is freely available, it is illegal in many sports, and has been implicated in many drug scandals.

Drugs have a tendency to mimic different neurotransmitters to different extents. For example, ecstasy is a strong serotonin mimic with some stimulant behaviour, but has a limited dopamine-like effect, so has a low addiction quality. Repeated use can cause overstimulation of neuroreceptors, resulting in serotonin burnout; symptoms including depression and other mood-related problems. This is because the brain becomes accustomed to large amounts of stimulus; consequently, regular levels of stimulation are perceived as being in deficit to what is required to feel 'normal'.

A compound called 3,4-Methyl-enedioxy-methamphetamine (MDMA), widely known as ecstasy, acts as a releasing agent of neurotransmitters, namely serotonin, norepinephrine and dopamine, which control mood and behaviour. These compounds are associated with release of the hormone oxytocin, which is implicated in the experiential qualities of the drug. Para-Methoxyamphetamine (PMA), like ecstasy, is serotonergic but takes longer to achieve the same effects. It may be sold instead of, or mixed with, MDMA because it is synthesised from more readily available starting materials (anisole) rather than the natural plant extract safrole, as with MDMA. What are the consequences of this? PMA is slower acting than MDMA; it takes longer than 30 minutes for any effect and 3 hours to reach maximum plasma levels; MDMA takes half this time, but the half-lives of the two are similar. Consequently, multiple dosing is common when the expected effects fail to occur quickly. Furthermore, the potency of PMA is 5–10 times that of MDMA, acting to both release serotonin and inhibit breakdown of the neurotransmitter, so low doses (plasma levels of 0.5 mg/L) cause serious elevated body temperature, resulting in overdose death. To put this in perspective, a typical 100 mg tablet of 'ecstasy' comprising 50 mg PMA will give a plasma PMA concentration of ca. 0.25 mg/L, i.e. two doses are likely to be fatal. Given that the user expects the rapid results of MDMA, when this fails to happen, multiple dosing will be a temptation.

The relatively simple structures of neuroactive drugs, combined with their straightforward synthesis and easy chemical manipulation, have led to the recent proliferation of psychoactive drugs which were effectively 'legal' because they have not been specifically outlawed. The most recent of these to gain widespread use and attention is mephedrone, which has since been made illegal. The structures of a selection of neuroactive drugs and the molecular structure of morphine are given in Figure 8.5.

The subtlety of neurotransmitter structure leads to the possibility of a number of side effects, and these are observed to greater and lesser extents for all drugs in this class. Neurotoxicity results from overstimulation of the neuroreceptors;

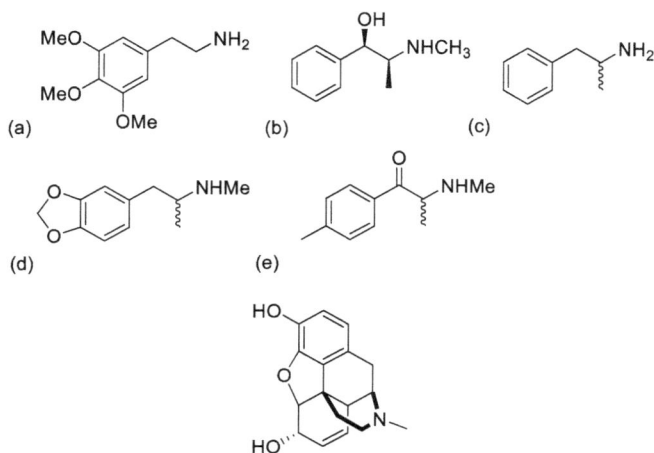

FIGURE 8.5 Structures of (a) mescaline, (b) ephedrine, (c) amphetamine, (d) ecstasy, (e) mephedrone and morphine (below).

the body becomes more adept at metabolising this type of compound to eliminate it, so eventually much higher doses are needed to achieve the desired response. Cardiotoxicity is another serious potential problem, which can result from stimulation of adrenal receptors in the heart. The extent of these problems is often unknown for these types of drugs because they have never gone through any clinical trials, instead going straight from preclinical development to mass human consumption and the potential hazards of this may not be known for a number of years.

Natural extracts from the poppy plant have been used as effective painkillers, as well as recreationally throughout history. Opiates are not the only compounds used for pain relief, for example, aspirin belongs to a different class of analgesics and works by different mechanisms to cure different types of pain. For a long time, medicinal chemists have searched for a non-addictive analgesic based on the opiate structure, but this has proven difficult. Opium alkaloids are the nitrogen-containing natural products extracted from the opium poppy, *Papaver somniferum*. These compounds are simplistic in structure and several thousands of these alkaloids have been isolated and characterised to give a huge library of biologically active compounds with potential as lead compounds in many fields of medicinal chemistry.

Historical uses of opiates have been recorded as early as 2000 years ago in ancient China, and because of its properties, the ancient Greeks dedicated the opium poppy to Thanatos (God of death), Hypnos (god of sleep) and Morpheus (god of dreams). Later physicians prescribed it for a range of afflictions, such as headache, epilepsy, vertigo and asthma, as well as use as a sedative. Use as an analgesic came much later. In the 16th century, Paracelsus introduced a preparation of opium called laudanum. The drug became increasingly popular, due to its wonderful effects. However, its addictive qualities led to problems which arose with long-term use; observations by doctors quoted 'great intolerable distress, anxieties and depression of the spirit'. Of course, these were symptoms of withdrawal.

Opiates were first marketed in Britain by Thomas Dover. Dover had a colourful past, with an early career in piracy and famed for rescuing the marooned sailor Alexander Selkirk from an uninhabited island, which is thought to be the inspiration for Dafoe's Robinson Crusoe. Dover later in life took up medicine and one of his products was a powder containing opium, liquorice, saltpetre and ipecacuanha, which is an emetic to make consumers sick if they took too much.

Godfrey's cordial from the 18th century was another popular remedy, containing opium, treacle and sassafras, used for treating rheumatic pains. This preparation was freely available in most grocery shops without prescription or restriction, despite many people becoming addicted. As late as the 20th century, opium was viewed as a legitimate substance, like tobacco and tea, but by 1920, the use of opiates was prohibited, and could only be prescribed by doctors if patients demonstrated a legitimate medical need.

Obsolescence of opium ensued after the raw material was superseded by purified, semi-synthetic and synthetic opiate analogues being developed. Morphine was the first pharmaceutical to be isolated from a plant extract. It is the main alkaloid present in opium and is responsible for analgesic activity. It was discovered in 1804 by Friedrich Wilhelm and Adam Sertürner, who after 13 years of research published the isolation of morphine in 1817. The advantage of the purified substance was that patients could be treated with a known dose and avoid the dangerous effects that were associated with consuming large quantities of opium. The sale of morphine commenced in 1827 by Heinrich Emanuel Merck of Darmstadt, who was able to expand his family pharmacy into the Merck KGaA pharmaceutical company. Heroin was the first semi-synthetic opioid and was marketed in 1898 by Bayer as a safe alternative to morphine because its lethal dose is hundreds of times above its effective dose. Morphine affects the central nervous system by binding to receptors in the brain and spinal cord, as well as to receptors present in the stomach and intestine, potentially causing harmful effects such as respiratory difficulties, coma or cardiac or respiratory collapse.

Targeting the nervous system can be a chancy business; medicinal chemists must carefully design drugs so that side effects are minimal, but create compounds with enough potency to be therapeutically useful. Due to the subtleties of the cholinergic and adrenergic receptors, greater attention is needed in understanding structure–activity relationships because active compounds will all have similar molecular structures, and small alterations to the structure may have dramatic consequences regarding activity. The reward for creating these drugs is substantial because disorders of the nervous system can be very problematic for patients. Unfortunately, the effects of psychoactive drugs can be desirable recreationally and this can lead to the abuse of certain compounds.

8.1.7 NEURODEGENERATIVE DISEASE

Beta-site amyloid precursor protein cleaving enzyme (BACE), also known as beta secretase, is an aspartic acid protease that is important in the formation of myelin sheaths in peripheral nerve cells. It is a transmembrane protein, which contains two active site aspartate residues in its extracellular protein domain and may function as a dimer. Elevated levels of this enzyme are present in patients with late-onset sporadic Alzheimer's disease. Generation of amyloid-β peptides, which aggregate in the brain of Alzheimer's patients, consists of 40 or 42 residues produced by two sequential cleavages of the amyloid precursor protein (APP) by BACE and gamma secretase. Initial cleavage by BACE creates a soluble extracellular fragment and a cell membrane-bound fragment, referred to as C99, which is cleaved in its transmembrane domain by gamma secretase to release amyloid-β in the cell. Initial cleavage of APP by α-secretase rather than BACE prevents the eventual generation of amyloid-β. The physiological purpose of BACE is unclear.

BACE inhibitors are a potential hit: drugs that block this enzyme, in theory, would prevent the build-up of amyloid-β and may help to slow or stop Alzheimer's disease. BACE1 is a compact globular integral membrane glycoprotein, comprising two domains: a cytoplasmic domain, where C-terminus cysteine residues serve as anchor points to the membrane, and a large ectodomain that contains the residues responsible for the catalytic activity. According to the crystal structure, the catalytic region is located between the N- and C-terminal lobes, within the substrate binding site in the cleft. The active site contains the two conserved aspartic acid residues, Asp32 and Asp228, within conserved motifs of eukaryotic aspartic proteases.

To test the effect of the drug, i.e. inhibiting BACE1, knockout mice were produced, with no biological deficiency, so specific inhibition of the enzyme was likely to be well tolerated, although there was a significant increase in neonatal lethality in mice. To find small molecule inhibitors of BACE1, fragment-based lead generation was applied. Fragment hits from a 1D NMR screening were characterised and the most promising compounds soaked into protein crystals to help facilitate a more rational design of more potent hit analogues to produce a library of heterocyclic amine compounds, with the potential for optimisation of drug-like properties.

Working on hits found for BACE1 inhibitors, it is necessary to introduce properties to the compounds that display the essential characteristics of a drug. In addition to peptide features and molecular size, there are several challenges that must be overcome in order to bring a drug-like BACE1 inhibitor to clinical development. Selectivity over other aspartic proteases is very important; with aspartic acid proteases being found in nearly all body tissues, off-target activity is possible and likely if an inhibitor is not specific to BACE1.

Molecular size was no trivial problem in the development of BACE1 inhibitors; the large size of the substrate pocket meant that, initially, decreasing the size of a peptide inhibitor resulted in a decrease in potency. Large inhibitors are not practical *in vivo*, therefore smaller inhibitors were needed, but they would not be able to fill the pocket as adequately as APP and thus were not good inhibitors. Some of this difficulty can be overcome by increasing the affinity of the inhibitor for the BACE1 active site through increasing interactions with the enzyme in the active site. Specific fragment compounds were found that formed strong hydrogen bonds to the critical Asp32 and Asp228 catalytic residues and development of a scaffold molecule that merged these fragments was accomplished.

Difficulties arose in producing a drug that would pass the blood–brain barrier. The inhibitors must penetrate the blood–brain barrier to exert their effect. Sterol linkers have been used to show cell membrane permeability as a limiting factor in BACE1 inhibition. Linking sterol to an inhibitor that was enzymatically potent greatly lost activity in cellular assays but enabled the inhibitor to permeate the cell membrane and reach the endosomes, where BACE1 is most active. P-glycoprotein efflux, which is the mechanism by which the inhibitor is removed from cells, is another limiting factor in the efficacy of BACE1 inhibitors.

Potential side effects that could arise from the drug have been indicated by test mice, suggesting that BACE proteases are necessary for the proper function of muscle spindles. It has been concluded that the drugs may have significant side effects related to impaired motor coordination. However, the BACE knockout mice were healthy.

BACE1 inhibitor AZD3839 was designed and developed at AstraZeneca and selected for clinical development. The lead structure (Figure 8.6) was identified by NMR-based fragment screening. Subsequent lead optimisation and design of new structural classes through scaffold hopping approaches led to a series of potent BACE1 inhibitors (2)–(4) with isoindole structural template. Fluorine can be introduced on the aromatic rings to manipulate the pharmacological properties. Introduction of difluoromethyl group on the pyridine in inhibitor (4) led to fluoroderivative (5) (AZD3839).

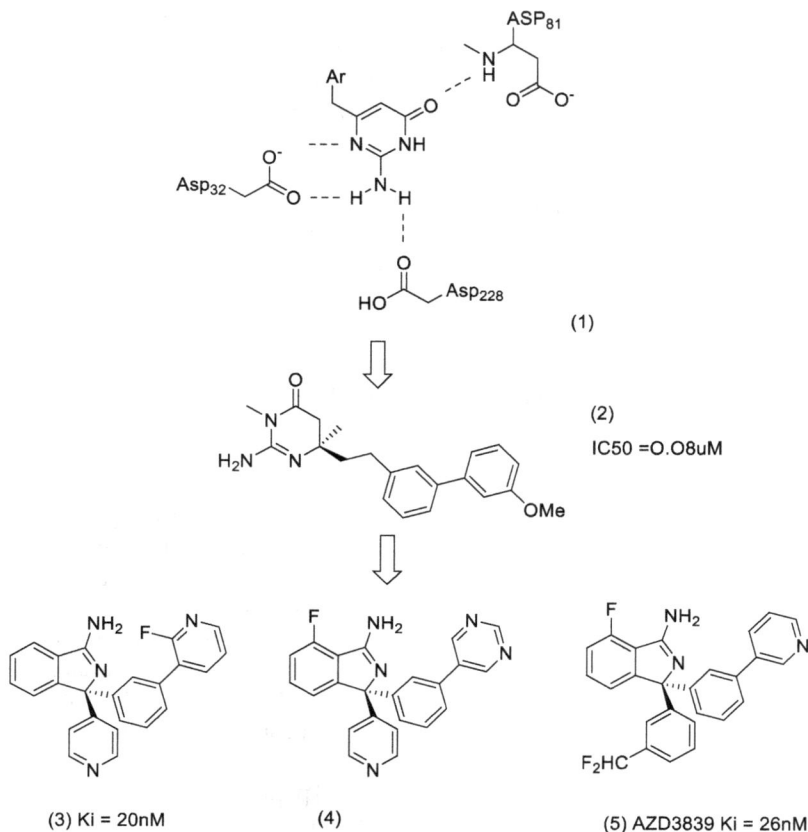

FIGURE 8.6 Route of development by AstraZeneca to optimise BACE inhibitor.

The binding mode of AZD3839 was determined by X-ray structural studies of inhibitor and BACE complex (Figure 8.7). It showed that the inhibitor binds in a flap-open conformation of the enzyme. The catalytic aspartates form a network of hydrogen bonds with the amidine group. The phenyl ring nestles in the S1 sub-pocket while the pyrimidine ring projects towards the S3 pocket. The substituted pyridine ring fills in the S2' sub-pocket and the pyridine nitrogen forms a hydrogen bond with Trp76[3].

8.2 DIABETES

In modern society, where people have become increasingly sedentary and have access to more sugary foods, problems with controlling blood sugar levels have become increasingly common and incidences of diabetes have risen dramatically. It is therefore important to understand how the metabolism works, the hormones and receptors involved, and contemplate how medicine can be used to help keep blood sugar levels in balance.

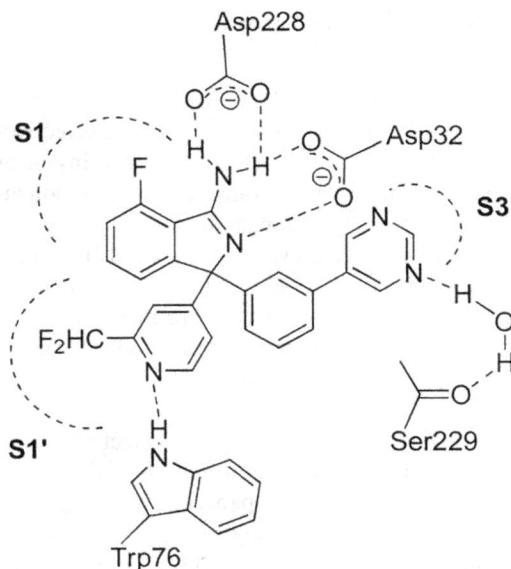

FIGURE 8.7   Drug interaction with active site.

8.2.1   Metabolism and the Generation of Energy

Understanding how the biochemistry of the body works and its nutritional demands is indispensable for someone engaged with any form of biomedical science in order to be able to appreciate the significance that nutrition has in terms of health and the medical consequences that poor nutrition can have. The food we consume is used to generate energy; these biochemical reactions are called metabolism. This energy is used to sustain muscle contractions, repair the body and support the immune system, as well as many other habituations of the body.

The energy that the body generates for these purposes is produced through respiration, a series of biochemical processes which derive chemical energy from the oxidation of carbohydrates. The body is efficiently designed to carry out respiration. Air is breathed in through the lungs, which are like a sponge, containing many small sacks called alveoli which increase the surface area of the lungs equivalent to the size of a football pitch to vastly increase the efficiency of oxygen diffusion from the air to the blood, via a network of very fine capillaries in the lungs. The oxygen is carried away in red blood cells; it binds to haemoglobin in the red blood cells, which is a globular protein composed of four subunits. Each one has an iron-containing haem group, enabling four oxygen molecules to be carried per haemoglobin molecule. The blood circulates around the body being pumped by the heart. The oxygen then dissociates from the haemoglobin when it reaches tissues that are in oxygen debt. Oxygen then diffuses into the cells of these tissues where it is utilised for aerobic respiration in order to generate the chemical energy ATP that the body requires to operate.

8.2.2 Homeostasis: The Body in Balance

The body must maintain a relatively constant internal environment by having control mechanisms that counteract external influences. This is known as homeostasis. For example, changes in body temperature would impact the effectiveness of the body's cells to conduct their routine operations due to the lowering of enzyme activity. If the body is cold, more heat must be generated by the metabolism of fuels in order for the muscles to work effectively. Changes in water potential cause the turgidity of cells to change due to osmosis, preventing the cells from operating as normal. Here, electrolytes are important to counteract this effect. The body utilises feedback loops to coordinate these control mechanisms. For these processes to operate properly though, it is important to pay attention to nutrition; enough calories need to be consumed to keep the body fuelled and warm and the body must be well hydrated with electrolytes.

Regulation of blood sugar levels is an essential aspect of homeostasis and instrumental in maintaining correct metabolism of fuel molecules. Control of blood sugar levels is performed by the pancreas. The pancreas is an organ made up largely of cells that are responsible for producing the enzymes: protease, amylase and lipase, required for digestion. Scattered throughout these cells though are the **islets of Langerhans**, which are groups of hormone-producing cells of which there are two types. The **α cells** are larger and produce the hormone glucagon. The **β cells** are smaller and produce the hormone insulin. These two hormones have antagonistic effects and are instrumental in regulating blood sugar levels.

Glucose is needed during respiration to produce ATP, which is the energy currency of cells. Restricted availability of glucose is problematic because energy is generated by alternative means, resulting in the production of ketone bodies. Normal blood glucose concentrations are 90 mg per 100 mL of blood. This glucose comes from three main sources. It can be obtained directly from the diet where carbohydrates are broken down during digestion; it can be mobilised from the breakdown of glycogen stores in the liver and muscles, which is called **glycogenolysis**; and it can be made from glycerol and amino acids, which is called **gluconeogenesis**.

Because food is eaten sporadically throughout the day at mealtimes and meals often vary in their content, the intake of glucose varies throughout the day. In addition, the demand for glucose varies, depending on levels of mental and physical activity; hence it is necessary to have hormonal control to maintain a constant blood glucose concentration.

The hormone **glucagon** has a role in raising blood glucose levels. The α cells of the islets of Langerhans detect a fall in blood glucose concentration and secrete glucagon directly into the blood plasma. Glucagon binds to receptor sites that occur on liver cell membranes, and this activates the enzymes within the cells that convert glycogen to glucose during glycogenolysis and also the conversion of glycerol and amino acids to glucose in gluconeogenesis. In muscle cells, breakdown of glycogen releases glucose, which is immediately available to fuel muscle contractions etc. Once blood glucose has risen, the α cells release less glucagon, which is a negative feedback loop.

The hormone **insulin** has a role in lowering blood glucose levels. The β cells of the islets of Langerhans detect a rise in blood glucose concentration and secrete insulin directly into the blood plasma. Most cells have glycoprotein receptors on their cell surface membranes to which insulin can bind. The attachment of insulin to the receptor site on glucose transport proteins causes the protein channel to open and glucose is able to diffuse into the cell. Additionally, the enzymes involved in the conversion of glucose into glycogen (glycogenesis) are activated. Overall, glucose levels in the blood are decreased in the following ways: increased absorption of glucose into cells, particularly muscle cells; increased respiration uses up more glucose; increased rate of glycogenesis in the liver and muscles; and increased rate of conversion of glucose to fat.

Blood glucose level, to an extent, is also regulated by adrenaline during the 'fight or flight' response, and this is known as the secondary messenger model and has two main effects: it activates glycogenolysis and deactivates glycogenesis. This process serves as an additional means of producing mobilised glucose for when demand is high. The adrenaline that is released during intense exercise or exertion therefore has a role in making glucose available to fuel the exertions.

Diabetes is a condition where regulation of blood sugar level is disrupted. In type 1 cases, the β cells do not produce enough insulin, consequence of an autoimmune response during early childhood development. After eating, glucose levels rise and remain high (**hyperglycaemia**), the kidneys cannot reabsorb all the glucose. This can lead to symptoms affecting vision, excessive urination, tiredness and weight loss. Control requires insulin injections. People with type 2 diabetes develop the condition later in life and is linked with obesity. Not enough insulin is produced by the β cells, and the glycoprotein cell receptors become insensitive to insulin, so lose their responsiveness. The treatment in this instance can be just to control the diet and exercise, or sometimes medication is also needed.

8.2.3 THE BIOCHEMISTRY OF METABOLISM AND THE GENERATION OF ENERGY

Metabolism is the collective network of biochemical reactions that occur within the body through which cells generate energy and reduce power to synthesise the essential molecules that the cells require to live. The astonishing array of chemical reactions involved in metabolism are in fact structured in a coherent manner with many common motifs, where the number of different kinds of reactions is comparatively small. Moreover, metabolic pathways are regulated in common ways: controlled in order to preserve homeostasis.

The laws of thermodynamics dictate that chemical reactions require a continual input of free energy. Cells need this energy for three main purposes: (1) the performance of mechanical work, such as muscle contraction; (2) the active transport of molecules and ions, which is critical for biochemical processes in the cell to occur; and (3) the synthesis of macromolecules and other biomolecules from simple precursors, so that the cell has the necessary materials needed to function properly. In biological systems, the fundamental currency of free energy is adenosine triphosphate (ATP). This chemical energy is obtained from the oxidation of foodstuffs during respiration. A large amount of free energy is liberated when ATP is hydrolysed to adenosine diphosphate (ADP) and inorganic phosphate, P_i, during a reversible process which is the fundamental mode of energy transfer in biological systems. This provides an immediate source of energy for cells rather than a long-term supply of energy.

The fuel required for generating the chemical energy ATP is obtained from foodstuffs. The first stage of the process is digestion. In the stomach and gut, hydrochloric acid, as well as a cocktail of digestive enzymes, break down the large molecules in food into smaller units. Proteins are hydrolysed to their constituent amino acids; polysaccharides are hydrolysed to their simple sugars, such as glucose; and fats are hydrolysed to glycerol and fatty acids. In the second stage, once these small molecules are assimilated (absorbed from the gut into the blood), they are converted into a few simple units that are key to metabolism. Lastly, these fuel molecules are taken up by cells and may be oxidised to produce ATP during the citric acid cycle and oxidative phosphorylation.

Carbohydrates are utilised as a primary source of energy from food. These are macromolecules, termed polysaccharides because they comprise long chains of sugar molecules linked together by glycosidic bonds. Carbohydrates serve as energy stores, fuel sources and metabolic intermediates, as well as having key roles in cellular recognition. Carbohydrates are present in many different forms. In plants, cellulose is used in cell walls to give rigidity and cannot be digested by humans. Starch, however, which makes up the energy reserves in plants and grains (potatoes, for example, are particularly rich in starch) consists of long linear chains of glucose molecules, which can be digested. Energy reserves in animals are stored as glycogen. These branched chains of glucose are very dense and can be liberated when blood sugar levels are lowered. Families of disaccharides are also abundant in many foods, consisting of bonded pairs of sugar molecules. Common disaccharides include sucrose: common table sugar from cane of sugar beet, lactose in milk and maltose from hydrolysis of starch. In each case, two sugar molecules are bonded by a simple glycosidic bond: glucose–fructose, galactose–glucose and glucose–glucose, respectively.

In digestion, glycosidic bonds are hydrolysed; the chemical bonds are broken to release water, and the monomer subunits can be assimilated. These can then be used for metabolic processes, for example, phosphorylated sugars are key intermediates in energy generation and biosynthesis. Hydrolysis occurs via specific enzymes. Sucrase, lactase and maltase, which are located in the epithelial cells lining the small intestine, specifically break down the disaccharides indicated from the nomenclature. The lining of the small intestine has many folds called villi and the epithelial cells contain many finger-like projections called microvilli and this serves to markedly increase the surface area for digestion.

Lactose intolerance occurs from being deficient in the enzyme lactase. Most adults are deficient in producing lactase; therefore, after ingestion of dairy, lactose sugar accumulates in the lumen of the small intestine because there is no mechanism for the uptake of this disaccharide from the gut. An influx of fluid into the small intestine results from the osmotic effect of the high concentration of sugars, which causes digestive discomfort and diarrhoea. A person's tolerance to dairy should therefore be considered when organising a diet plan/supplement regime (whey is a dairy product, although lactose is often removed) with regard to digestive requirements.

Animal cells store glucose in the form of glycogen, which is a highly branched polymer of glucose residues. The branching serves to increase the solubility of glycogen and make its sugar units more readily mobilised. The greatest stores of glycogen are found in skeletal muscle and the liver. Plants store starch as their nutritional reservoir. There are two types: amylose is an unbranched form of starch and amylopectin is the branched form, albeit not as highly branched as glycogen. More than half the carbohydrate ingested by humans is starch. Both forms of starch are hydrolysed by the enzyme amylase, produced in the saliva and in the pancreas for digestion. Starch is broken down to maltose, maltotriose and dextrin (made of several glucose units) which are then further degraded to their glucose subunits by the relevant enzymes, so that the glucose can be made ready for cellular processes.

Respiration is the primary source of glucose consumption, required for the generation of metabolic energy. Respiration is a complex biochemical process, which involves several discrete metabolic reactions and specialised structures within the cell in order to conduct these processes. Glycolysis is the first stage of respiration and in aerobic systems is the prelude to the citric acid cycle and ETC (electron transport chain) whereby ATP is synthesised in the cell's mitochondria.

In terms of a chemical definition of the process, glycolysis is the sequence of reactions that converts glucose into pyruvate with the concomitant production of ATP. Pyruvate is a biochemical intermediate which then enters the mitochondria from the cell cytoplasm, where, provided sufficient oxygen, it is further degraded during the citric acid cycle and ETC, which are the major processes for harvesting energy from glucose. However, with inadequate oxygen supplies, as in the case of actively contracting muscles, pyruvate is converted into lactic acid, bypassing the citric acid cycle and ETC.

The energy economy of oxidative phosphorylation versus glycolysis illustrates the efficiency of aerobic respiration versus anaerobic respiration and explains why high levels of physical activity cannot be sustained. During the processes of glycolysis, the citric acid cycle and fatty acid oxidation, molecules of NADH and $FADH_2$ are formed. These are energy-rich molecules as they contain a free pair of electrons with transfer potential, which are donated to molecular oxygen to yield large amounts of free energy, used to generate ATP during the ETC. Oxidative phosphorylation is the process by which ATP is formed as electrons are transferred from NADH or $FADH_2$ to O_2 by a series of electron carriers in the mitochondrial membrane. Oxidative phosphorylation generates 32 of the 36 ATP molecules that are formed when glucose is completely oxidised to CO_2 and H_2O.

The mitochondria are oval-shaped organelles that are effectively the batteries of the cell, specially designed for generating metabolic energy. The mitochondria consist of inner and outer membranes where ATP is made and an internal matrix, where the citric acid cycle and fatty acid oxidation occur to make NADH and $FADH_2$ which are utilised in oxidative phosphorylation.

The body has a high demand for glucose: around 160 g is required daily, 120 g of which is needed for the brain alone. Glucose can also be synthesised from non-carbohydrate precursors, which is vital to maintain homeostasis because the daily intake of glucose fluctuates. This process is called gluconeogenesis, which becomes important in periods of fasting and periods of intense exercise. To meet metabolic demands, gluconeogenesis occurs in the liver and kidneys to help maintain glucose levels in the blood so that the brain and muscle have sufficient quantities available to function normally. The gluconeogenesis pathway converts pyruvate into glucose. The pyruvate is produced mainly from lactic acid that is produced in the active skeletal muscles when the rate of glycolysis exceeds the metabolic rate of the citric acid cycle and respiratory chain under conditions of oxygen deficit.

Most of the body's readily available glucose is stored as glycogen. The large, branched polymer of glucose is broken down by enzymes during glycogen metabolism to mobilise the glucose and make it readily available in the blood between meals and for periods of strenuous muscular activity. Indeed, in an average 70 kg man, the glucose in body fluids has an energy content of only 40 kcal, whereas the energy content of stored glycogen is greater than 600 kcal, even after an overnight fast. Most of this glycogen is stored in the liver and muscles, and present as granules in the cell cytosol.

Turning attention away from carbohydrates, fatty acids are also important biomolecules. They are the main component of cell membranes, having endocrine properties, and fatty acids are an important source of fuel. They are stored as triacylglycerols, which are esters of glycerol made of hydrocarbon chains with a terminal carboxyl group. The chain length and degree of unsaturation (number of double C=C in the chain) vary a great deal. This has implications regarding the metabolism of fatty acids, with saturated fats being more difficult to break down.

The triacylglycerols are highly concentrated stores of metabolic energy. The complete oxidation of fatty acids will yield about 9 kcal/g compared to about 4 kcal/g for carbohydrates and proteins. Fat metabolism has a greater caloric yield because triacylglycerols are more highly reduced than carbohydrates or proteins and they are anhydrous, so they do not bind water molecules and therefore have a much greater energy density. The fatty acids are stored in specialised cells called adipose tissue (fat cells) contained as globules which can account for most of the cell's volume. These cells are specialised for the synthesis and storage of triacylglycerols and for their mobilisation as fuel molecules transported in the blood. This activity is regulated by hormones.

Fat metabolism occurs via a catabolic process known as beta oxidation, during which fatty acid molecules are broken down in the mitochondria for use in respiration to generate metabolic energy. The fatty acids are broken down into acetyl-CoA, which enters the citric acid cycle, and NADH and $FADH_2$, which are coenzymes used in the electron transport chain. Exercise increases fat oxidation by upregulating the enzymes involved in beta oxidation of fatty acids. Exercise also increases blood flow to muscles, where the fat is used for energy generation. Over the longer term, regular vigorous exercise has the effect of increasing mitochondrial mass and the network of blood vessel to supply muscles with oxygenated blood.

8.2.4 ENERGY SYSTEMS

Oxygen is taken up via the respiratory system and transported by haemoglobin along the cardiovascular system to the cells of the muscle where it is needed for the generation of energy. Energy production occurs in the cell mitochondria, where respiration produces chemical energy, ATP. ATP is made up of adenosine bonded to three phosphate molecules. When these strong covalent bonds are broken to remove a phosphate group, energy is released.

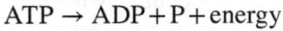

ATP $\rightarrow$ ADP + P + energy

There is only a limited supply of ATP stored in the body, enough for 1–2 seconds of activity, so this molecule has to be synthesised from fats and carbohydrates stored in the body as glycogen via the process of respiration. There are three primary systems for regenerating ATP from ADP and inorganic phosphate.

- Creatine phosphate (CP) system – creatine phosphate donates a phosphate ion to generate energy. Heavy weight training and sprinting emphasise the CP system, as energy is needed to be generated quickly. However, this is not sustainable and the muscles soon fatigue.
- Lactate system – energy is produced from glucose during respiration in the absence of oxygen. This occurs during glycolysis and the Krebs cycle and produces lactic acid as a by-product, which gives the burning sensation in fatigued muscle.
- Aerobic (oxidative) system – energy is produced from carbohydrates and fats in the presence of oxygen during moderately intense exercise, such as jogging. This process is called aerobic respiration.

8.2.5 OTHER IMPORTANT HORMONES

Besides the hormones used to regulate blood sugar and energy generation, there are many other aspects of the endocrine system that have a strong influence on metabolism. Sex hormones, testosterone and oestrogen, are responsible for secondary sexual characteristics. However, testosterone plays an important role in muscle growth and repair in both males and females. It is produced in the testes of males and in small amounts (about 1/10th that of males) in the ovaries and adrenal of females.

Testosterone is associated with anabolism, the repair of tissues. Cortisol is the hormone involved in catabolism, the breakdown of tissues. Under times of stress, such as during exercise, cortisol is secreted from the adrenal glands to maintain energy supply by the breakdown of carbohydrates, fats and protein. High levels of cortisol caused by overtraining, poor sleep, poor nutrition or high stress can lead to significant levels of muscle breakdown.

Growth hormone is released from the pituitary gland in the brain and is regulated by the nearby hypothalamus. Growth hormone may be stimulated by testosterone, oestrogen, deep sleep and vigorous exercise. It is mainly an anabolic hormone, important for growth during childhood and plays a role in bone development and muscle protein synthesis in adulthood.

The thyroid gland is located at the base of the neck, and thyroid hormones are associated with metabolism and are regulated by the pituitary gland. Thyroid hormones play a key role in basal metabolism, the metabolism of macronutrients, adrenaline sensitivity, protein synthesis, as well as heart rate, breathing rate and body temperature regulation.

8.2.6 MEDICATION FOR DIABETES

Insulin is important for the regulation of blood sugar levels. It is essential for people with type 1 diabetes, who cannot control their blood sugar levels, but may not be necessary for people with type 2 diabetes. Most people require medicine to control their type 2 diabetes. This helps to keep the patients' blood sugar level as normal as possible to prevent health problems associated with the disease. If identified early, type 2 diabetes can be managed through diet and lifestyle. However, many people have to take diabetic medication for the rest of their life, although the particular medicine or dose may need to change over time. Medication is ideally used in conjunction with a healthier lifestyle. Adjusting diet and being active are usually also necessary to keep blood sugar levels down.

There are many types of medicine for type 2 diabetes. It can take time to find a medicine and dose that's right for any given patient. Hence, people with diabetes will often have to go through a period of trial to find which medication works best for them, subject to regular review to ensure that the condition is managed as well as possible. Usually, people will be offered a medicine called metformin first. It may be required to take extra medicines, or a different medicine such as insulin.

If metformin does not work well enough on its own, or the patient cannot take it or they have other health problems, patients can be offered other medicines alongside or instead of metformin. These include other tablets that help lower your blood sugar, such as gliclazide, glimepiride, alogliptin, sitagliptin or pioglitazone. Tablets that lower blood sugar and help your heart pump blood around your body are those such as dapagliflozin, empagliflozin, ertugliflozin or canagliflozin. Injections or tablets that lower blood sugar and help you lose weight are those such as semaglutide, dulaglutide or tirzepatide.

People with diabetes will need insulin if other medicines no longer work well enough to keep blood sugar below your target. For those with type 1 diabetes, insulin is essential. Sometimes type 2 diabetics may need insulin for a short time, such as if they are pregnant, if they are ill or to bring blood sugar level down when first diagnosed. It is necessary to inject insulin using an insulin pen. This is a device that helps the patient to inject safely and take the right dose.

8.3 THE CARDIOVASCULAR SYSTEM AND HEART DISEASE

Oxygen from the air is breathed into the lungs and travels via the respiratory system and is taken to the muscles and organs, where it is needed, by the cardiac system, as illustrated in Figure 8.8. Blood is the universal transport system of the body. An average adult will have 4.5–5.5 L of blood, which carries oxygen and a plethora of other substances around the body. The blood plasma is mostly water and contains red blood cells, white blood cells and platelets. Red blood cells contain haemoglobin, which is the protein that binds oxygen and carries it to the muscles and organs. White blood cells are responsible for fighting infection and are an essential part of the immune system. Platelets facilitate blood clotting to prevent hazardous amounts of blood loss from an injury. Of course, the blood is pumped around the body by the heart. The heart is made of specialised cardiac muscle. It is evolved to perform its

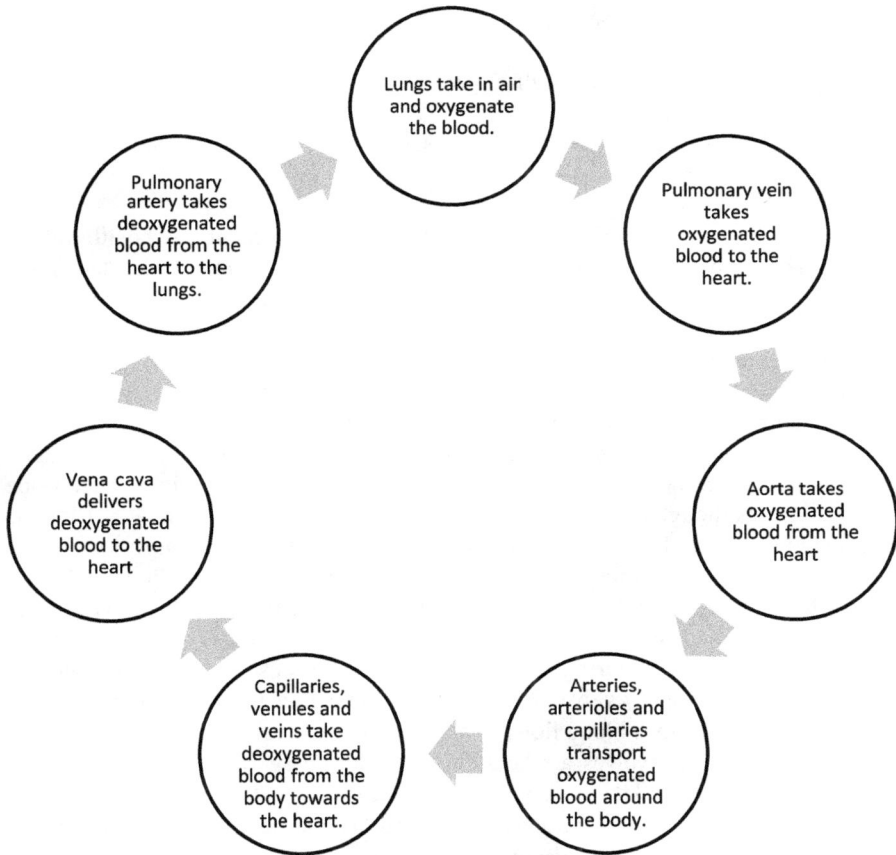

FIGURE 8.8 Flow diagram of blood transport via the cardiovascular system.

function without fatiguing, which would obviously be disastrous. The circulatory system comprises the heart and blood vessels which form a closed continuous loop that serves the entire body.

Arteries are thick and muscular in order to transport blood away from the heart under high pressure. Their elastic property allows them to expand as the heart beats. Veins take blood back towards the heart, so are under considerably less pressure. Hence the veins have a less thick layer of smooth muscle and contain one-way valves to prevent blood flowing in the wrong direction. Varicose veins arise when these valves do not function properly, leading to blood pooling. Smaller arteries and veins are referred to as arterioles and venules, respectively. The smallest blood vessels that branch out from the circulatory system and deliver blood to remote parts of the body are called capillaries. They are one cell thick and permeable to allow efficient diffusion of substances. Physical exercise will lead to an increase in capillarisation, as the body adapts to supply more blood to the muscles.

- Blood pressure is a measure of force that the blood exerts on the walls of the arteries.
- Systolic pressure is the (higher) pressure in the arterial system when the heart is beating.
- Diastolic pressure is the (lower) pressure in the arterial system when the heart is refilling.

This is why blood pressure readings are given as measurements. A healthy blood pressure should be in the range of 80–120/60–80 and can vary a little, depending on multiple factors. Some of the factors that affect blood pressure include cardiac output, peripheral vascular resistance (dependent on the elasticity of the vascular system) and the viscosity of blood.

The heart is a fist-sized, four-chambered muscular pump, located slightly left of central, behind the sternum. The left-hand side of the heart receives and pumps out oxygenated blood, while the right-hand side receives and pumps out deoxygenated blood. Diagrams of the heart (Figure 8.9) are transposed so that the right and left side are reversed, as though describing the heart of someone that is facing you. Blood is received in the atria, then passes to the ventricles where their thick muscle wall contracts to vent blood from the heart. Atrioventricular valves prevent the backflow of blood when the heart muscle contracts. On the left side of the heart, there is a mitral (bicuspid) valve and on the right-hand side of the heart there is a tricuspid valve. These prevent blood re-entering the atria from the ventricles. There are also semilunar valves; the pulmonary valve prevents blood flowing back into the right ventricle and the aortic valve stops blood flowing back into the left ventricle.

The cardiac cycle involves a relaxation (diastole) phase where relaxation of the atria allows blood to refill them from the pulmonary veins and vena cava. Meanwhile, blood flows into the relaxed ventricles. Atrial systole (contraction) pushes the remaining blood into the ventricles (stimulated by the sinoatrial node). Then, ventricular systole (contraction) pressurises the blood, closing the atrioventricular valves and

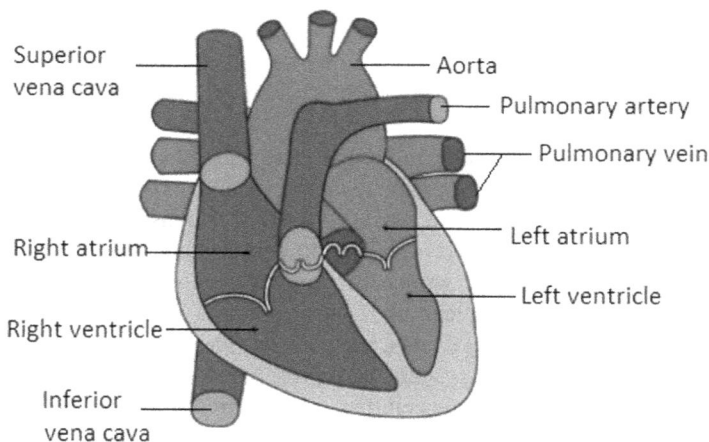

FIGURE 8.9   Schematic diagram of the heart.

forces blood from the heart via the pulmonary artery and aorta. This is stimulated by the bundle of His, which is proceeded by another phase of relaxation. On average, the heart will beat like this 72 times in a minute while the body is at rest.

8.3.1 CARDIOVASCULAR MEDICINE

Coronary heart disease (CHD) is usually caused by a build-up of fatty deposits called **atheroma** on the walls of the arteries around the heart (coronary arteries). The build-up of atheroma makes the arteries narrower, restricting the flow of blood to the heart muscle. This process is called atherosclerosis.

A person's risk of developing atherosclerosis is significantly increased if they: smoke, have high blood pressure (hypertension), have high cholesterol, have high levels of lipoprotein (a), do not exercise regularly or have diabetes. Other risk factors for developing atherosclerosis include being obese or overweight, or having a family history of CHD – the risk is increased if a person has a male relative under the age of 55, or a female relative under 65, with CHD.

Smoking is a major risk factor for coronary heart disease. Both nicotine and carbon monoxide (from the smoke) put a strain on the heart by making it work faster. They also increase a person's risk of developing blood clots. Other chemicals in cigarette smoke can damage the lining of the coronary arteries, leading to furring of the arteries.

High blood pressure (hypertension) puts a strain on the heart and can lead to CHD. Hypertension can also be caused by high cholesterol, and in turn, increase the risk of CHD. Cholesterol is a fat made by the liver from the saturated fat in your diet. It's essential for healthy cells, but too much in the blood can lead to CHD. Like cholesterol, lipoprotein (a), also known as Lp(a), is a type of fat made by the liver. It's a known risk factor for cardiovascular disease and atherosclerosis. The level of Lp(a) in the blood is inherited from a person's parents. It's not routinely measured, but screening is recommended for people with a moderate or high risk of developing cardiovascular disease.

Lifestyle is important in mitigating the risk of developing CHD. Besides smoking and poor diet, lack of exercise is another critical factor. If a person is often inactive, fatty deposits can build up in their arteries. If the arteries that supply blood to the heart become blocked, it can lead to a heart attack. If the arteries that supply blood to the brain are affected it can cause a stroke.

A high blood sugar level may lead to diabetes, which can more than double the risk of developing CHD. Diabetes can lead to CHD because it may cause the lining of blood vessels to become thicker, which can restrict blood flow. All these factors can lead to thrombosis. A **thrombosis** is a blood clot in a vein or artery. If a thrombosis develops in a coronary artery, it prevents the blood supply from reaching the heart muscle. This usually leads to a heart attack.

8.3.2 TREATMENT OF HEART DISEASE

Treatment for coronary heart disease (CHD) can help manage the symptoms and reduce the risk of further problems. CHD can be managed effectively with a combination of lifestyle changes and medicine; however, in some cases, surgery may be required. With the correct treatment, the symptoms of CHD can be reduced, and the functioning of the heart improved.

In the first instance, people diagnosed with coronary heart disease, by making simple lifestyle changes, can reduce their risk of having further episodes. For example, stopping smoking after a heart attack quickly reduces the risk of having a heart attack in the future close to that of a non-smoker. Other lifestyle changes, such as having a healthy diet and doing regular exercise, will also reduce the risk of heart disease in the future.

Many different medicines are used to treat CHD. Usually, they either aim to reduce blood pressure or widen the arteries. **Blood thinners** are a type of medicine that can help reduce the risk of a heart attack by thinning the blood and preventing it clotting. This reduces the risk of thrombosis, where a blood clot can block an artery supplying the heart with blood. Common blood-thinning medicines include: low-dose aspirin, clopidogrel, rivaroxaban, ticagrelor, and prasugrel.

People with high cholesterol are at a great risk of developing CHD, therefore cholesterol-lowering medicine called **statins** may be prescribed to help minimise the risk. Examples include: atorvastatin, simvastatin, rosuvastatin, pravastatin. Statins work by blocking the formation of cholesterol and increasing the number of low-density lipoprotein (LDL) receptors in the liver. This helps remove LDL cholesterol from the blood, which makes a heart attack less likely. However, the suitability of statins for individual patients at risk of CHD needs to be assessed by a GP. In some cases, beta-blockers may be more appropriate.

Beta-blockers, including atenolol, bisoprolol, metoprolol and nebivolol, are often used to prevent **angina** and treat high blood pressure. They work by blocking the effects of hormones, such as adrenaline in the body, which slows down a person's heart rate and improves blood flow.

Vasodilators, such as nitrates, are used to widen the blood vessels. They are available in a variety of forms, including tablets, sprays and skin patches such as glyceryl trinitrate and isosorbide mononitrate. Nitrates work by relaxing the blood vessels, which therefore widen, letting more blood pass through them. This lowers blood pressure and hopefully relieves any heart pain that the patient may have. Nitrates can have some mild side effects, including headaches, dizziness and flushed skin. **Calcium channel blockers** also work to decrease blood pressure by relaxing the muscles that make up the walls of arteries. This causes the arteries to become wider, reducing your blood pressure. Examples include amlodipine, verapamil and diltiazem.

Angiotensin-converting enzyme (ACE) inhibitors are commonly used to treat high blood pressure. Examples include ramipril and lisinopril. They block the activity of a hormone called angiotensin 2, which causes the blood vessels to narrow. The resulting widening of the blood vessels makes it easier for the heart to pump blood around the body. In addition to stopping the heart working as hard, ACE inhibitors improve the flow of blood around the body. Due to the effects of these medicines, and how they work, patients' blood pressure will have to be monitored while taking ACE inhibitors, and regular blood tests will be needed to check that your kidneys are working properly. However, less than 1 in 100 people have problems with the blood supply to their kidneys (renal stenosis) as a result of taking ACE inhibitors.

Side effects of ACE inhibitors can include a dry cough and dizziness. **Angiotensin-2 receptor blockers (ARBs)** have a similar effect to ACE inhibitors. They are used to lower blood pressure by blocking angiotensin-2. Mild dizziness is usually the only side effect, so they are often prescribed as an alternative to ACE inhibitors, as they do not cause a dry cough. **Diuretics**, which are commonly known as water pills, work by flushing excess water and salt from the body through urine. This can help to relieve pressure on the heart muscle.

Procedures and surgery are sometimes necessary, in chronic cases of CHD. If a person's blood vessels are narrow as the result of a build-up of **atheroma** (fatty deposits) or if symptoms cannot be controlled using medicines, interventional procedures or surgery may be needed to open up or bypass blocked arteries.

TUTORIAL QUESTIONS

1. Briefly describe how N-alkylated analogues of morphine could be synthesised.
2. Explain how the tricyclic antidepressants work at controlling mood.
3. Catecholamines can be developed into anti-asthmatic drugs. Outline which groups on catecholamines are important for binding and explain what changes could be made to the structure of catecholamines to make them selective towards β-receptors.
4. Explain how ramipril and lisinopril are used to reduce the risk of coronary heart disease.
5. Why are statins given to people at risk of heart disease?

Conclusion

Ever since the discovery of the atom through Rutherford's alpha-scattering experiments in 1911, our knowledge and understanding of chemistry has catapulted the development of the subject, which has been instrumental in facilitating many advancements in our modern society. The nature of atoms and understanding how they interact and bond with one another is fundamental to chemistry. Having an appreciation of the three-dimensional properties of molecules is essential to structure and bonding; molecular geometry and isomerism are important features of drug molecules and govern how drug molecules bind to their target through intermolecular interactions.

To design and synthesise drugs, the interactions of molecules in terms of organic reactions and their relevance to synthesis are foundational to medicinal chemistry. Computational modelling has the capability to reveal potential molecular structures which might interact with a biological receptor in a way that is medicinally beneficial. The challenge for organic chemists is then to use their knowledge of the subject to design a strategy to synthesise the target molecule (TM) from readily available staring materials (RASMs), which ideally are compounds that are inexpensive and commercially manufactured. In general, many similar structures are synthesised in what is referred to as a compound library. Structures produced as potential hits are screened and those compounds found to have good structure–activity relationships go on for lead optimisation.

In order to design a target molecule for a biological receptor, medicinal chemists need to understand the molecular basis of diseases to be able to identify targets. Targets are often proteins; hence an understanding of protein structure is important. Knowledge of the amino acid sequence and the three-dimensional tertiary or quaternary structures is vital for building a model to investigate possible binding interactions of ligands, which could potentially lead to drug development. Their importance in catalysis, cellular recognition and structure means that proteins have a high likelihood of being possible targets for a disease. It is also important to understand genetics because genes contain the underlying information for protein synthesis, and often the mechanisms of a disease; therefore they present other potential targets such as gene therapy. Indeed, the future of medicinal chemistry is likely to be based more around genetics as our understanding of the subject grows.

Many of the first cases where drugs were rationally designed against identified targets were in the treatment of viral diseases. Targets were revealed from molecular modelling, based on information from X-ray crystallography of virus components, then molecular structures could be designed that would be expected to interact with the virus in a certain way. This is a good example of where understanding the molecular basis of disease is utilised in medicinal chemistry, for example the influenza life cycle, which was discovered from structural analysis and viral genetic sequencing. Viral genetic sequencing yields the potential for synthesis of pure proteins that can be used for drug–target binding assessments.

Drugs can be designed to operate on targets in different ways, for example, inhibition of an enzyme by a compound that mimics the natural substrate to block the active site, such as in the case of Relenza™ and Tamiflu™ blocking the neuraminidase active site, an enzyme important in the life cycle of the influenza virus. Compounds can be designed to interact with a target with subtle properties of the molecule deliberated during optimisation to have the effect of agonism or antagonism, depending on what property is required. By understanding the protein 3D structure–function relationship and building the drug molecule appropriately, accounting for hydrogen bonding and optimising all non-covalent interactions, a drug can be designed to either enhance or suppress, respectively, the activity that results from the natural substrate. The subtle results for correct optimisation were exemplified by Relenza™ and Tamiflu™ in terms of the method of administration. Although both structures achieved strong binding that was needed for inhibition of neuraminidase, Tamiflu was designed with more hydrophobic moieties which meant that it could be absorbed across the gut and was orally active, which is preferable to Relenza which had to be inhaled, which could lead to irritation.

Rational design of drugs to an identified target can often take the approach of combinatorial chemistry, known as the molecular fragments technique, where sections of a compound that are discovered to bind well to a section of the target are spatially positioned and a scaffold section of the molecule is designed to hold the fragments in these positions. This is how GlaxoSmithKline and Roche came to develop similar structures independently. The next stage is lead optimisation, where drug properties are optimised, for example, the improved oral bioavailability of Tamiflu compared to Relenza. Idealised drug criteria for an effective medicine involve: high affinity for the target, safe/well tolerated by the patient, synergistic with other drugs when used in combination therapy and can be taken orally with minimal dosing frequency. The closer a drug adheres to these requirements, the greater its potential as a medicine.

Interestingly, Tamiflu is prepared through a complicated 15-step synthesis, involving shikimic acid, extracted from star anise as a starting material. Despite the lengthy synthesis, Roche uses 90% of the global crop of star anise to make Tamiflu because of the mass market and profitability from government stockpiling of the drug. Unfortunately, though, some influenza strains now show resistance to Tamiflu. The mechanism of mutation involves a single amino acid mutation in the neuraminidase active site which prevents effective drug binding without loss of function of the enzyme.

The occurrence of resistance and its challenges to medicine are highly relevant to antibiotics and the treatment of bacterial infections. Indeed, resistance to antibiotics probably has greater media attention than for any other class of drugs and this is because it is such an important issue. To demonstrate how crucial antibiotics have been to developing society in the past, consider the morbidity in history prior to antibiotics. The advent of antibiotics was a cultural and medical revolution, resulting in increased success rates of surgery and a dramatic increase in life expectancy.

As the field of medicine advances into the future, it is vital to recognise the challenges of today: problems in developing countries with antibiotic availability, cholera, epidemics and the risk from globalisation regarding the transmission pathogens. More frequently, pathogenic diseases that were once eradicated from the developed world are reappearing during the modern era of mass migration of people. The occurrence of Antibiotic resistance in the developed world, with notable examples such as multidrug-resistant tuberculosis (TB) and MRSA, are a public health crisis. As we move into the future, medicinal chemistry will need innovation, designing new drugs to battle infection.

Paramount for designing new drugs, an understanding of the molecular basis of the disease, as emphasised so many times throughout this text, is essential for understanding the resistance mechanisms that bacteria employ. Bacteria can produce more of an enzyme that metabolises the drug, change the target enzyme active site subtly, so that the drug cannot bind effectively and therefore does not work, and decrease cell wall permeability to the drug or express more efflux pumps to remove the drug from the cell; all of these mechanisms derive from mutation. An example case study is penicillin resistance, where mutation led to the production of an enzyme, β-lactamase, which broke down the drug. Consequently, strategies needed to be employed to combat this problem, such as the use of clavulanic acid administered as Augmentin, to improve the efficacy of amoxicillin. Chemists need to identify targets to address these problems. Strategies adopted in the past such as steric shields to prevent access to the β-lactamase active site, for example methicillin, relied on understanding the binding interaction with the target. However, steric shields must not compromise enzyme–substrate complementarity with the target. Another approach was to search for other targets and develop drugs with different mechanisms of action, for example, vancomycin acted on the bacterial cell wall via a different mechanism to penicillin. Finding alternative targets besides the bacterial cell wall was another strategy, such as in the case of tetracycline, which acts to disrupt bacterial protein synthesis.

Resistance to vancomycin is a big problem because this drug is often regarded as the last line of defence against pathogenic bacteria. 'Superbugs' as pathogenic bacteria become tolerant to a number of antibiotics, resulting in multidrug-resistant genes on plasmids that can be rapidly transferred between bacteria and to daughter cells, which is a huge challenge to medicinal chemistry. Resistance can arise form overuse of antibiotics, people not completing a course of medication, trace amounts of antibiotics in hospitals, which are all influencing factors resulting in the creation of superbugs that develop tolerance to many kinds of antibiotic. These are all habits that need to be addressed to help prolong the life expectancy of antibiotics. In the future, there could be the possibility of identifying new targets from genome sequencing and understanding the genetic-level mechanisms of disease. This could allow for the development of a whole new class of antibiotics that pathogens are not familiar with.

Keeping one step ahead of mutation will be key to combating pathogens in the future, for example, utilising a multidrug approach to avoid selection of resistant pathogens like the case of the influenza virus varying the NA binding site, or in the battle against MRSA. It is likely that the arms race between pathogens and drugs will never cease. The unnecessary use of drugs should undoubtedly be avoided to try and preserve the arsenal we have against pathogens for as long as possible. However, from the individual patient's perspective, being denied medicine when they feel unwell and the need to justify their access to drugs can be a challenge, and restricting drug use can also impact the profitability of pharma industry, which is critical for research and development. These ethical and philosophical considerations could affect the evolution of medicine in the future.

One of the greatest medical challenges to modern society is cancer. With a growing and ageing population, the occurrence of the disease has skyrocketed, with one in three people developing the disease according to the American Cancer Society. Cancer can be a consequence of external influences such as exposure to chemicals that are damaging to DNA, carcinogens and viruses which carry oncogenes. These lead to mutations of DNA and if genes controlling cell cycle are damaged, this results in uncontrolled replication of cells into tumours. This can be an age-related breakdown of control mechanisms as well as longer exposure to carcinogens. Cancer can also spread, which is known as metastasis. This is another challenge of the disease, making it harder to control.

A molecular understanding of DNA replication helps to understand the disease and assists to find ways to combat cancer. One possibility might be to look at mitosis to find targets. Understanding the molecular basis of complex diseases to identify targets and develop therapies, for example, from understanding mechanisms of the cell cycle and DNA replication, might elucidate ways in which the disease might be treated, but the high complexity means that this is an immense task. As cancer treatment advances into the future, it will be vital to find mechanism-specific ways to combat the disease because the current approaches to cancer therapy have so many problems. Traditional ways to target cancer have harsh side effects; targeting DNA directly in the case of alkylating agents and intercalators, which act by distorting the α-helix and cross-link strands to inhibit DNA replication, also effects normal cells. Mustard cross-linking agents and cisplatin cross-linking agents both have highly adverse side effects because of their influence on normal cells as well. Use of metal complexes has been explored to produce less toxic analogues of cisplatin, but these still cause harm to normal cells. Hormone therapy, targeting the oestrogen receptor, is another method to fight cancer, but the lack of discrimination for normal cells still remains.

Frontiers in cancer treatment look to create therapies without the severe side effects of traditional chemotherapy. The methods focus more on target specificity, such as the use of antibodies in antibody-directed enzyme prodrug therapy (ADEPT). This technique harnesses a 'seek and destroy' approach where the specificity of antibodies is utilised to take the drug directly to the site of action, thereby minimising the effect on normal cells. Ingenious strategies such as this will be the focus of cancer therapy in the future.

Neurodegenerative diseases, which are associated with old age, will become increasingly common in the future with ageing populations. Conditions such as Alzheimer's disease result from the breakdown of normal biochemical processes in the neurons of the brain. Beta-site amyloid precursor protein cleaving enzyme (BACE), also known as beta secretase, is an aspartic acid protease that is important in the formation of myelin sheaths in peripheral nerve cells. Elevated levels of this enzyme are present in patients with late-onset sporadic Alzheimer's disease. Amyloid-β peptides, which aggregate in the brain of Alzheimer's patients to form amyloid plaques, are formed from the amyloid precursor protein (APP) after cleavage by BACE. Initial cleavage of APP by α-secretase rather than BACE prevents the eventual generation of amyloid-β, which would otherwise form the plaque which causes impaired brain function and the symptoms of Alzheimer's. Currently, research focused on finding inhibitors of BACE to stop the formation of amyloid plaque is ongoing. The physiological purpose of BACE is unclear but will be a focus of Alzheimer's research in the future.

To conclude, in order to overcome many of the challenges that medicine faces in the future, medical research will be reliant on the work of medicinal chemists to develop drugs that will help to treat the problems caused by diseases that current therapies fail to address. These solutions, just like the pioneering work on drug discovery in the past, will be governed by the fundamental principles of chemistry. Foundational to organic chemistry, the construct of all living organisms is based on the element carbon. Atoms of this element are bonded together by sharing pairs of electrons to build the scaffold onto which other elements essential to organic chemistry, most notably hydrogen, oxygen and nitrogen, can bond to create the macromolecules, such as proteins, nucleic acids, carbohydrates and lipids, which comprise the materials for building the cell. Biochemists study cells and the molecules of which they are composed to understand how cellular processes operate and relate to the function of the body, including the occurrence of disease. By understanding the molecular basis of disease, chemists are able to intelligently design drugs that will influence the biology of the cell in a way that alleviates the symptoms of the disease. The product of a century of scientific ingenuity, medicines are available today that would have seemed miraculous 100 years ago. It is worth acknowledging though that the challenges presented to us by diseases in the future will require advancements in the field of medicinal chemistry that will be perceived as equally extraordinary.

Index

Note: **Bold** page numbers refer to tables and *italic* page numbers refer to figures.

For Product Safety Concerns and Information please contact our EU
representative GPSR@taylorandfrancis.com
Taylor & Francis Verlag GmbH, Kaufingerstraße 24, 80331 München, Germany

www.ingramcontent.com/pod-product-compliance
Lightning Source LLC
Chambersburg PA
CBHW060553220326
41598CB00024B/3086